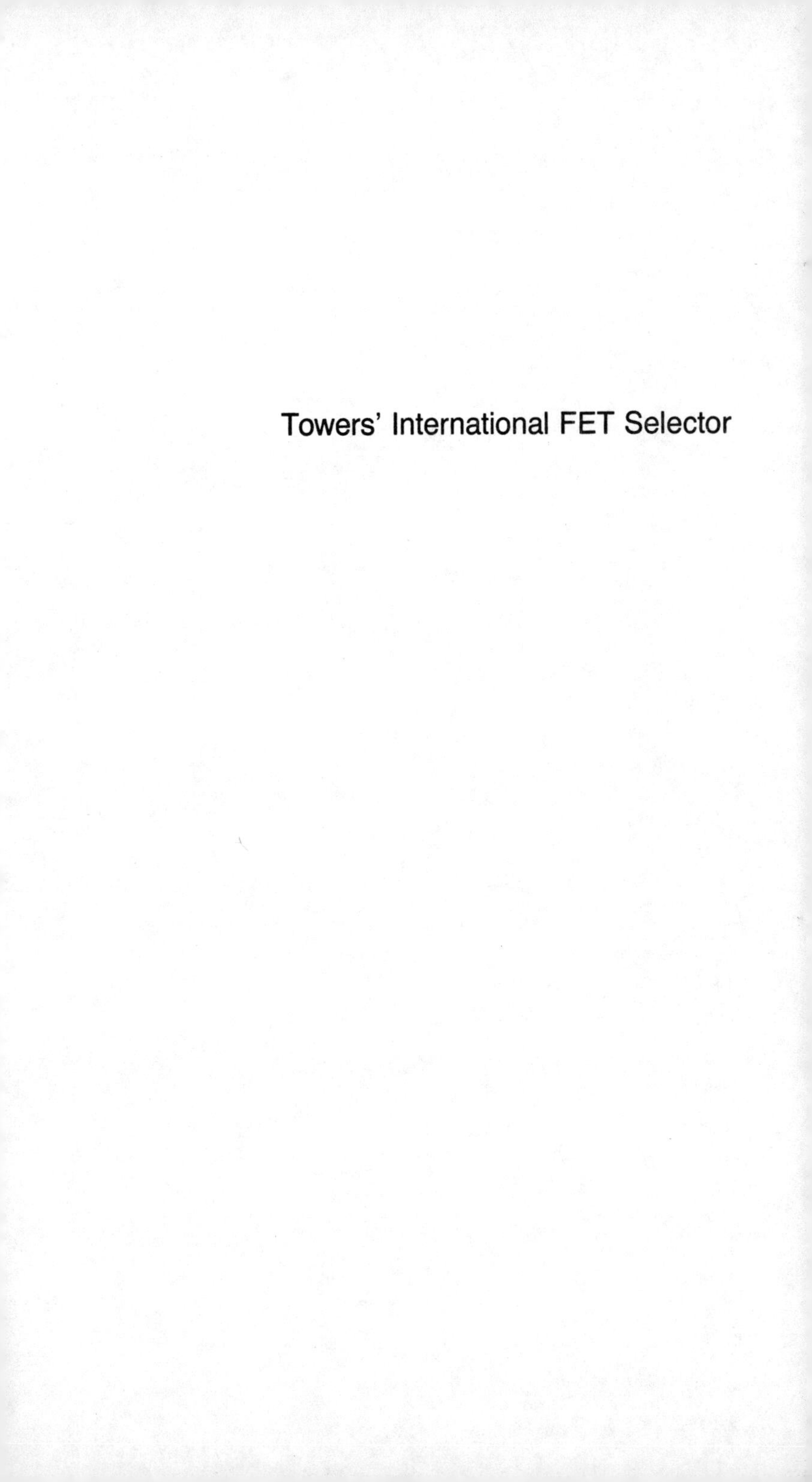

Towers' International FET Selector

Towers' International FET Selector

Specification data for the identification, selection and substitution of FETs

by T D Towers, MBE, MA, BSc, C Eng, MIERE
and N S Towers, BA (Cantab).

FIRST PRINTING—JANUARY 1978

Originally published by
W. FOULSHAM & CO. LTD., England

Library of Congress Cataloging in Publication Data

Towers, T. D., 1914-
Tower's International FET selector.

Includes index.
1. Field-effect transistors--Handbooks, manuals, etc.
I. Title. II. Title: International FET selector.
TK7871.95.T67 621.3815'28 77-18894
ISBN 0-8306-9988-0
ISBN 0-8306-1016-2 pbk.

Preface

If you deal with field effect transistors, or FETs—whether as a student, a hobbyist, a circuit engineer, a buyer, a teacher or a serviceman—you often want data on a specific FET for which you know only the type number. Or, you may be even more interested in where you can get the device in question. And perhaps more important still (particularly with obsolete devices), you may want guidance on a readily available substitute.

This FET compendium, a comprehensive tabulation of basic specifications, offers information on:

1. Ratings
2. Characteristics
3. Case details
4. Terminal identifications
5. Applications use
6. Manufacturers
7. Substitution equivalents (both European and American)

The many FETs covered in this compendium are most of the more common current and widely-used obsolete types.

It is international in scope and covers FETs not only from the USA and Continental Europe, but also from the United Kingdom and Japan.

Contents

An Introduction to the FET

The FET (Field Effect Transistor) is a semiconductor amplifying device, which, along with the bipolar transistor, has largely replaced the vacuum tube in electronics.

Called Unipolar Transistor and Tecnetron in the 1950s, the FET did not come into general use until the late 1960s when silicon manufacturing techniques, developed for ordinary transistors, made mass production of FETs possible.

Like the bipolar transistor, the FET is a three-terminal amplifying device which comes in two polarities for use with either + or − supply voltage. It is packaged in conventional, small, flexible-lead packages. Where the ordinary transistor requires a current *into* its input (base) terminal to control a larger current at its output (between emitter and collector terminals), the FET uses a voltage on the gate (base) terminal to control the current between its source (emitter) and drain (collector). Thus, where a bipolar transistor gain is characterized as current gain, the FET gain is characterized as a transconductance, or mutual conductance (g_m)—the ratio of change in output (drain) current to input (gate) voltage. The gain g_m is usually expressed either in milliamps per volt (mA/V) or micromhos (μmho), where 100 μmho = 1 mA/V.

One of the most important differences in the FET from the bipolar transistor is in its input impedance. In the transistor, the base input corresponds to a forward-biased diode with an impedance of typically hundreds or thousands of ohms. In the FET the input impedance corresponds to a reverse-biased diode which typically exhibits tens of megohms input impedance.

Originally known as unipolar transistors, FETs are governed by only one type of internal current carrier, either the hole or the electron, depending on the device polarity. Ordinary transistors were called bipolar because the device currents are conducted by two types of carrier, both electrons and holes, whatever the device polarity.

FETs come in two polarities: n-channel, corresponding to npn transistors (+ supply) and p-channel, corresponding to pnp transistors (– supply).

Next, FETs can be junction-type (with the gate input being a diode) or insulated-gate (with the gate input being an insulating oxide layer). Common abbreviations are JFET for junction gate types and MOS for insulated-gate types. Also the insulated-gate (IGFET) is commonly termed MOSFET (for Metal Oxide Silicon FET).

Finally, FETs come in two operational modes: depletion and enhancement. Depletion FETs pass maximum current when the input is grounded (gate connected to source), and are normally operated with reverse bias on the gate to reduce the bias current below the maximum (like a tube). Enhancement FETs pass no current when the input is grounded, and have to be operated with the input gate forward biased to set up the required bias current. Occasionally depletion-enhancement types are found where a bias current flows with the input grounded which can then be reduced by reverse biasing the input or increased by forward biasing. FETs are depletion type only, but MOSFETs can be depletion, depletion-enhancement or enhancement.

Many different circuit symbols have been used over the years for FETs. A selection of the more common are set out below for information. The main points to note are that arrow points on the left hand gate lead indicates junction types, and lack of gate arrow indicates insulated-gate types. In addition, a solid cross-bar in the symbol indicates depletion types and a broken one enhancement types.

When you are considering an FET type, the first thing to do is confirm whether it is n-channel (by far the commonest) or p-channel. Any replacement must obviously be of the same polarity.

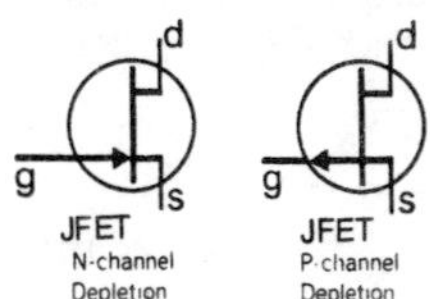

Except for a few isolated germanium types, FETs are normally manufactured from silicon, and, other things equal, the basic material of the device is not important.

Many differently shaped cases have been developed over the years for FETs, and you will need to know details of the case outline in any FET you are going to use. This selector provides such details, along with lead/terminal indications.

The FET is basically a relatively low voltage device, and it is important to check the permissible maximum voltage that can be applied to it. This selector gives output voltage ratings (drain-source) since this is the critical characteristic that mainly concerns the ordinary user.

Tubes have one big advantage over FETs. They can withstand substantial current overloads for short periods without catastrophic results. FETs are catastrophically sensitive to current overloads. It is important, therefore, to know the maximum permissible drain current, recorded in the selector, and ensure that this is not exceded under any circumstances.

The FET is not as sensitive to heat overloads as it is to voltage or current. The junction heats up inside the FET under load and manufacturers carefully specify maximum permissible junction temperature to guard against excessive degradation of device characteristics with time. Permissible maximum junction temperature for FETs (silicon types) usually lie between 125° C and 200° C.

For power dissipation, manufacturers specify either a free air rating at 25° C for low power devices normally mounted by their leads only, or a case-rated dissipation at 25° C case temperature for high power devices usually mounted on some form of heat sink. Occasionally, however, the maximum dissipation is specified as attached to a heatsink at 25 ° C ambient temperature. It is important to realize that the permissible dissipation in use must be reduced, i.e. the device derated, as the ambient temperature rises above 25° C. For a device rated P_{TOT} at 25° C and with a maximum permissible junction temperature of T_j, the permissible dissipation at an elevated ambient temperature of T_a can be shown to be $P_{TOT} = P_{TOT} (T_j - T_a)/(T_j - 25)$.

One of the most important characteristics of the FET is its transconductance (mutual conductance), g_m or g_{fs}: the ratio of the output drain current to the input gate voltage. It is necessary to know the transconductance not only to design the DC bias networks for the device but to compute the stage gain for amplification. FET transconductance changes rapidly with the level of standing drain bias current, but it is relatively insensitive to drain standing bias voltage. This selector quotes the transconductance at maximum current.

FET formulas exist for computing working parameters from limit values given as follows:

$I_D = I_{DSS} (1 - V_G/V_P)^2$,
$g_m = g_{mo} (1 - V_G/V_P)$,
$g_m^2 = g_{mo}^2 I_D/I_{DSS}$,
$g_{mo} V_P = 21_{DSS}$

Finally the high frequency characteristics of the device are important. The two that give the greatest measure of useful information are C_{iss} (gate input capacitance) and C_{rss}(feedback capacitance). Generally, any replacement should be a device with similar C_{iss} and C_{rss} if used at high frequencies.

The selector tables give general application guidance on the use of each FET by a special three-letter coding in a separate column.

Additionally, in the last two columns the tables offer possible substitute devices. Substitution guidance given should be used with caution because special characteristics not recorded may make the suggested substitute unacceptable for some applications. You should get detailed specification of the proposed substitute from the manufacturer and consider them in relation to your own application before committing yourself. The suggested substitutes can be taken only as a general guide.

Over the years, manufacturers and standards associations have issued over two thousand separate FET numbers. Some of these were never widely used or are no longer used by designers.

Devices either have type numbers peculiar to individual manufacturers (see *Appendix C*) or conform to industry-standard serial numbering systems (suppliable by more than one manufacturer).

The principal industry standard numbering systems are:

(a) **2N.** The EIA in the USA maintains a register of "**2N**" types familiarly known as "JEDEC" types, which have had world wide acceptance. In the JEDEC numbering, FETs may appear under **2N** (for 3-lead types) or **3N** (for 4-lead), but the serial numbers give no indication of polarity, etc.

(b) **Proelectron.** The Association Internationale Pro Electron in Europe maintains a register of Pro Electron types which have wide acceptance in Europe. FETs in the Pro Electron system are registered like bipolar transistors either under a two letter/three numeral code (e.g. BF244) for consumer devices, or a three letter/two numeral code (e.g. BFW11) for industrial devices. The second and third letters in the number have applications significance as indicated in *Appendix C*.

(c) **2S**. In Japan, the JIS, Japanese Industrial Standards, system of transistor numbering is almost universal, there being very few housecode devices. In this system, all FET numbers start with **2S** (for 3-lead) or **3S** (for 4-lead), followed by a letter and several numbers. The letter after the "S" has significance: J = FET, p-channel; K = FET, n-channel.

The FETs included in this selector are listed in serial numero-alphabetical order in tables immediately following this introduction. The tables are designed as far as possible to be self explanatory. For readers to whom some of the nomenclature is strange, there is in *Appendix A* an explanatory set of notes supplementing the tabulation. Besides detailed comments on the tables, *Appendix A* incorporates a ready reference chart of the tabular format and a glossary of terms used.

The FET tables are also supported by:

Appendix B: Package Outline and Lead/Terminal Diagrams.
Appendix C: Manufacturers' House Codes
Appendix D: Manufacturer Listings.

Every reasonable care has been taken to ensure accuracy of information in the tables, but no responsibility can be accepted for inaccuracies that may have arisen. The user who wishes fuller technical information on current devices should consult the data issued by the manufacturer.

Tabulations

TYPE NO	CONS TRUC TION	CASE & LEAD	V DS MAX	I D MAX	T J MAX	P TOT MAX	VP OR VT	IDSS OR IDOM	G MO	R DS MAX	C ISS MAX	C RSS MAX	USE	SUPP LIER	EUR SUB	USA SUB	ISS
1G2	NJD	B15	50V	16MA	200C	0.3WF	6.5MXV	16MXMA	4MNMO		7P5		ALG	OBS	BFW61	2N3822	0
2G2	NJD	B15	50V	8MA	200C	0.3WF	4MXV	8MXMA	2MNMO		7P5		ALG	OBS	BFW12	2N3822	0
2N2386	PJD	B21	20V	20MA	200C	0.5WF	8MXV	0.9/9MA	1MNMO		50P		ALG	TDY	BF320	2N2386A	0
2N2386A	PJD	B21	20V	15MA	200C	0.5WF	8MXV	1/15MA	2.2/5MO		18P		ALG	TDY	BF320	2N2386	0
2N2497	PJD	B21	20V	3MA	200C	0.5WF	5MXV	1/3MA	1/2MO		32P		ALG	TDY	BF320A	2N3329	0
2N2497JAN	PJD	B21	20V	10MA	200C	0.5WF	5MXV	3MXMA	1/2MO	1K	32P		RLS	TDY		2N2497	0
2N2498	PJD	B21	20V	6MA	200C	0.5WF	6MXV	2/6MA	1.5/3MO		32P		ALG	TDY	BF320B	2N3330	0
2N2498JAN	PJD	B21	20V	10MA	200C	0.5WF	6MXV	6MXMA	1.5/3MO	800R	32P		ALG	TDY		2N2498	0
2N2499	PJD	B21	20V	15MA	200C	0.5WF	8MXV	5/15MA	2/4MO		32P		ALG	TDY	BF320	2N2386A	0
2N2499A	PJD	B21	20V	15MA	200C	0.5WF	10MXV	5/15MA	1.7TPMO		32P		ALG	TDY	BF320	2N3909	0
2N2499JAN	PJD	B21	20V	25MA	200C	0.5WF	8MXV	15MXMA	2/4MO	600R	32P		RLS	TDY		2N2498	0
2N2500	PJD	B21	20V	6MA	200C	0.5WF	6MXV	1/6MA	1/2.2MO		32P		ALG	TDY	BF320B	2N3332	0
2N2500JAN	PJD	B21	20V	10MA	200C	0.5WF	6MXV	6MXMA	1/2.2MO		32P		RLS	TDY		2N2499	0
2N2606	PJD	B7	30V	0.5MA	175C	0.3WF	1/4V	.1/0.5MA	0.1MNMO		6P		ALG	INB	BF320A	2N2843	0
2N2606JAN	PJD	B7	30V		200C	0.3WF	4MXV	0.5MXMA	0.11/.6MO		6P		ALG	INB	BF320A	2N2606	0
2N2607	PJD	B7	30V	1.5MA	175C	0.3WF	1/4V	.3/1.5MA	0.33MNMO		10P		ALG	INB	BF320A	2N2844	0
2N2607JAN	PJD	B7	30V		200C	0.3WF	4MXV	1.5MXMA	0.33/1MO		10P		ALG	INB	BF320A	2N2607	0
2N2608	PJD	B7	30V	4.5MA	175C	0.3WF	1/4V	.9/4.5MA	1MNMO		17P		ALG	INB	BF320A	2N5460	0
2N2608CHP	PJD	B73	30V	10MA	200C		1/4V	.9/4.5MA	1MNMO		17P		ALG	SIU			0
2N2608JAN	PJD	B7	30V		200C	0.3WF	4MXV	4.5MXMA	1/2.5MO		17P		ALG	INB	BF320B	2N2608	0
2N2609	PJD	B7	30V	10MA	175C	0.3WF	1/4V	2/10MA	2.5MNMO		30P		ALG	INB	BF320	2N5464	0
2N2609CHP	PJD	B73	30V	10MA	200C		1/4V	2/10MA	2.5MNMO		30P		ALG	SIU			0
2N2609JAN	PJD	B7	30V		200C	0.3WF	4MXV	10MXMA	2.5/6.2MO		30P		ALG	SIU	BF320B	2N2609	0
2N2620	NJD	B47	50V	0.2A	175C	0.3WF		1.5/5MA	0.1MNMO				ALG	OBS	BFW12	2N3822	0
2N2794	PJD	B23	20V	5MA	175C	0.4WF		1.5/5MA	0.1MNMO				ALG	OBS	BF320A	2N3820	0

TYPE NO	CONS TRUC TION	CASE & LEAD	V DS MAX	I D MAX	T J MAX	P TOT MAX	VP OR VT	IDSS OR IDOM	G MO	R DS MAX	C ISS MAX	C RSS MAX	USE	SUPP LIER	EUR SUB	USA SUB	ISS
2N2841	PJD	B7	30V	2.2MA	200C	0.3WF	1.7MXV	.125MXMA	0.06MNMO		6P		RLS	SIU		2N3573	0
2N2842	PJD	B7	30V	2.2MA	200C	0.3WF	1.7MXV	.325MXMA	0.18MNMO		10P		RLS	SIU		2N3574	0
2N2843	PJD	B7	30V	2.2MA	200C	0.3WF	1.7MXV	0.2/1MA	0.54MNMO		17P		RLS	SIU		2N5265	0
2N2844	PJD	B7	30V	2.2MA	200C	0.3WF	1.7MXV	.4/2.2MA	1.4MNMO		30P		RLS	SIU	BF320A	2N2608	0
2N3066	NJD	B6	50V	4MA	200C	0.3WF	9.5MXV	0.8/4MA	0.4/1MO		10P	1P5	RLS	SIU	BF800	2N4302	0
2N3066A	NJD	B6	50V	4MA	200C	0.3WF	9.5MXV	0.8/4MA	0.4/1MO		10P	1P5	ALN	SIU	BF800	2N4302	0
2N3067	NJD	B6	50V	4MA	200C	0.3WF	4.5MXV	0.2/1MA	0.3/1MO		10P	1P5	ALG	SIU	BF800	2N3686A	0
2N3067A	NJD	B6	50V	4MA	200C	0.3WF	4.5MXV	0.2/1MA	0.3/1MO		10P	1P5	ALG	SIU	BF800	2N3686A	0
2N3068	NJD	B6	50V	4MA	200C	0.3WF	2.2MXV	0.25MXMA	0.2/1MO		10P	1P5	ALG	SIU	BF800	2N3687A	0
2N3068A	NJD	B6	50V	4MA	200C	0.3WF	2.2MXV	0.25MXMA	0.2/1MO		10P	1P5	ALG	SIU	BF800	2N3687A	0
2N3069	NJD	B6	50V	10MA	200C	.35WF	9.5MXV	2/10MA	1/2.5MO		15P	1P5	ALG	SIU	BFW12	2N3822	0
2N3069A	NJD	B6	50V	10MA	200C	.35WF	9.5MXV	2/10MA	1/2.5MO		15P	1P5	ALN	SIU	BFW56	2N3685A	0
2N3070	NJD	B6	50V	10MA	200C	.35WF	4.5MXV	.5/2.5MA	.75/2.5MO		15P	1P5	ALG	SIU	BF808	2N3685A	0
2N3070A	NJD	B6	50V	10MA	200C	.35WF	4.5MXV	.5/2.5MA	.75/2.5MO		15P	1P5	ALN	SIU	BF808	2N3685A	0
2N3071	NJD	B6	50V	10MA	200C	.35WF	2.2MXV	0.1/.6MA	0.5/2.5MO		15P	1P5	ALG	SIU	BF347	2N3687A	0
2N3084	NJD	B23	30V	3MA	200C	0.4WF	10MXV	3MXMA	0.4/2MO		14P		ALG	OBS	BF808	2N3685	0
2N3085	NJD	B47	30V	3MA	200C	0.4WF	10MXV	3MXMA	0.4/2MO		14P		ALG	CRY	BF808	2N3821	0
2N3086	NJD	B23	40V	3MA	200C	0.4WF	10MXV	3MXMA	0.4/2MO		14P		ALG	OBS	BF808	2N3685	0
2N3087	NJD	B47	40V	3MA	200C	0.4WF	10MXV	3MXMA	0.4/2MO		14P		ALG	CRY	BF808	2N3821	0
2N3088	NJD	B23	15V	3MA	200C	0.4WF	5MXV	2MXMA	0.4/4MO		14P		ALG	OBS	BF808	2N4339	0
2N3088A	NJD	B23	15V	3MA	200C	0.4WF	5MXV	2MXMA	0.4/2MO		14P		ALG	OBS	BF808A	2N3821	0
2N3089	NJD	B47	30V	2MA	200C	0.4WF	1/5V	0.5/2MA	0.3/2MO		6P	2P	ALG	SIU	BF800	2N3687A	0
2N3089A	NJD	B47	30V	2MA	200C	0.4WF	1/5V	0.5/2MA	0.3/2MO		6P	2P	ALN	SIU	BF200	2N3687A	0
2N3112	PJD	B7	20V	10MA	200C	0.3WF	4MXV	.18MXMA	.05/.12MO		3P5		ALG	SIU		2N2606	0
2N3113	PJD	B7	20V	1MA	200C	0.3WF	4MXV	0.18MXMA	0.05MNMO		2P		ALG	SIU		2N3573	0

TYPE NO	CONS TRUC TION	CASE & LEAD	V DS MAX	I D MAX	T J MAX	P TOT MAX	VP OR VT	IDSS OR IDOM	G MO	R DS MAX	C ISS MAX	C RSS MAX	USE	SUPP LIER	EUR SUB	USA SUB	ISS
2N3277	PJD	B62	25V	10MA	200C	0.4WF	5MX	.15/.5MA	0.1MNMO		4P5		ALG	INB		2N2606	0
2N3278	PJD	B62	25V	10MA	200C	0.4WF	5MXV	.5/.9MA	0.15MNMO		4P5		ALG	INB		2N5265	0
2N3328	PJD	B12	20V	1MA	200C	0.3WF	6MXV	1MXMA	0.1MNMO		4P		ALG	SIU	BF320A	2N2606	0
2N3329	PJD	B12	20V	15MA	200C	0.3WF	5MXV	1/3MA	1/2MO	1K	20P		ALN	SIU	BF320A	2N5266	0
2N3329CHP	PJD	B73	20V	3MA	200C		5MXV	1/3MA	1/2MO	1K	20P		ALG	SIU			0
2N3329JAN	PJD	B12	20V	10MA	200C	0.3WF	5MXV	3MXMA	1/2MO	1K	20P		ALG	SIU	BF320A	2N3329	0
2N3330	PJD	B12	20V	15MA	200C	0.3WF	6MXV	2/6MA	1.5/3MO	800R	20P		ALN	SIU	BF320B	2N3378	0
2N3330CHP	PJD	B73	20V	6MA	200C		6MXV	2/6MA	1.5/3MO	800R	20P		ALG	SIU			0
2N3330JAN	PJD	B12	20V	10MA	200C	0.3WF	6MXV	6MXMA	1.5/3MO	800R	20P		ALG	SIU	BF320B	2N3330	0
2N3331	PJD	B12	20V	15MA	200C	0.3WF	8MXV	5/15MA	2/4MO	600R	20P		ALN	SIU	BF320C	2N5462	0
2N3331CHP	PJD	B73	20V	15MA	200C		8MXV	5/15MA	2/4MO	600R	20P		ALG	SIU			0
2N3331JAN	PJD	B12	20V	10MA	200C	0.3WF	8MXV	15MXMA	2/4MO	600R	20P		ALG	SIU	BF320	2N3331	0
2N3332	PJD	B12	20V	15MA	200C	0.3WF	6MXV	1/6MA	1/2.2MO		20P		ALN	SIU	BF320B	2N2500	0
2N3332JAN	PJD	B12	20V	10MA	200C	0.3WF	6MXV	6MXMA	1/2.2MO	800R	20P		ALG	SIU	BF320B	2N3332	0
2N3333	PJD	B76	20V	10MA	200C	0.4WF	1.6MXV	1MXMA	0.6/1.8MO		30P		DUA	TIB			0
2N3334	PJD	B76	20V	10MA	200C	0.4WF	1.6MXV	1MXMA	0.6/1.8MO		30P		DUA	TIB			0
2N3335	PJD	B76	20V	10MA	200C	0.4WF	1.6MXMA	1MXMA	0.6/1.8MO		30P		DUA	TIB			0
2N3336	PJD	B76	20V	10MA	200C	0.4WF	1.6MXV	1MXMA	0.6/1.8MO		30P		DUA	TIB			0
2N3365	NJD	B6	40V	4MA	150C	0.3WF	11.5MXV	0.8/4MA	0.4/2M		15P	2P5	ALG	SIU	BF808	2N3685A	0
2N3366	NJD	B6	40V	4MA	150C	0.3WF	6.5MXV	0.2/1MA	0.25/1MO		15P	2P5	ALG	SIU	BF800	2N3687A	0
2N3367	NJD	B6	40V	4MA	150C	0.3WF	2.2MXV	0.25MXMA	0.1/1MO		15P	2P5	ALG	SIU	BF800	2N3687A	0
2N3368	NJD	B6	40V	12MA	150C	0.3WF	11.5MXV	2/12MA	1/4MO		20P	3P	ALG	SIU	BFW61	2N5458	0
2N3369	NJD	B6	40V	12MA	150C	0.3WF	6.5MXV	2.5MXMA	0.6/2.5MO		20P	3P	ALG	SIU	BF347	2N3821	0
2N3370	NJD	B6	40V	12MA	150C	0.3WF	3.2MXV	0.1/.6MA	0.3/2.5MO		20P	3P	ALG	SIU	BF800	2N3687A	0
2N3376	PJD	B12	30V	50MA	175C	0.3WF	1/5V	0.6/6MA	0.8/2.3MO	1K5	10P		RLS	SIU	BF320B	2N3378	0

TYPE NO	CONS TRUC TION	CASE & LEAD	V DS MAX	I D MAX	T J MAX	P TOT MAX	VP OR VT	IDSS OR IDOM	G MO	R DS MAX	C ISS MAX	C RSS MAX	USE	SUPP LIER	EUR SUB	USA SUB	ISS
2N3376CHP	PJD	B73	30V	6MA	200C		1/5V	0.6/6MA	0.8/2.3MO	1K5	6P		ALG	SIU			0
2N3377	PJD	B77	30V	100MA	200C	.15WF	5MXV	6MXMA	0.8/2.3MO	1K5	2P		RLS	OBS	BF320A	2N3376	0
2N3378	PJD	B12	30V	50MA	175C	0.3WF	4/5V	3/6MA	1.5/2.3MO	750R	10P		RLS	SIU	BF320B	2N5269	0
2N3378CHP	PJD	B73	30V	6MA	200C		4/5V	3/6MA	1.5/2.3MO	750R	6P		ALG	SIU			0
2N3379	PJD	B77	30V	100MA	200C	.15WF	5MXV	6MXMA	1.5/2.3MO	750R	2P		RLS	OBS	BF320A	2N3378	0
2N3380	PJD	B12	30V	50MA	175C	0.3WF	4/9.5V	3/20MA	1.5/3MO	600R	10P		RLS	SIU	BF320C	2N3331	0
2N3380CHP	PJD	B73	30V	20MA	200C		4/9.5V	3/20MA	1.5/3MO	600R	6P		ALG	SIU			0
2N3381	PJD	B77	30V	100MA	200C	.15WF	9.5MXV	20MXMA	1.5/3MO	600R	2P		RLS	OBS	BFT11	2N3380	0
2N3382	PJD	B12	30V	50MA	175C	0.3WF	1/5V	3/30MA	4.5/12MO	300R	30P		RLS	SIU	BFT11	2N3993	0
2N3383	PJD	B77	30V	100MA	200C	.15WF	5MXV	30MXMA	4.5/13MO	300R	5P		RLS	OBS	BFT11	2N3382	0
2N3384	PJD	B12	30V	50MA	175C	0.3WF	4/5V	15/30MA	7.5/12MO	180R	30P		RLS	SIU		2N5116	0
2N3385	PJD	B77	30V	100MA	200C	.15WF	5MXV	30MXMA	7.5/13MO	180R	5P		RLS	OBS	BFT11	2N3384	0
2N3386	PJD	B12	30V	50MA	175C	0.3WF	4/9.5V	15/50MA	7.5/15MO	150R	30P		RLS	SIU		2N5115	0
2N3387	PJD	B77	30V	100MA	200C	.15WF	9.5MXV	50MXMA	7.5/15MO	150R	7P		RLS	SIU		2N3386	0
2N3436	NJD	B6	50V	15MA	200C	0.3WF	9.8MXV	3/15MA	2.5/10MO		18P	6P	ALG	SIU	BFW61	2N4303	0
2N3437	NJD	B6	50V	15MA	200C	0.3WF	4.8MXV	0.8/4MA	1.5/6MO		18P	6P	ALG	SIU	BFW12	2N4302	0
2N3438	NJD	B6	50V	15MA	200C	0.3WF	2.3MXV	0.2/1MA	0.8/4.5MO		18P	6P	ALG	SIU	BFW13	2N3821	0
2N3452	NJD	B15	50V	4MA	150C	0.3WF	9.8MXV	0.8/4MA	0.2/1.2MO		6P	1P5	ALN	SIU	BF800	2N3685A	0
2N3453	NJD	B15	50V	4MA	150C	0.3WF	4.8MXV	0.2/1MA	0.15/.9MO		6P	1P5	ALN	SIU	BF800	2N3686A	0
2N3454	NJD	B15	50V	4MA	150C	0.3WF	2.3MXV	0.25MXMA	0.1/.6MO		6P	1P5	ALN	SIU	BF800	2N3687A	0
2N3455	NJD	B15	50V	4MA	150C	0.3WF	9.8MXV	0.8/4MA	0.4/1.2MO		5P	1P5	ALG	SIU	BF800	2N4867A	0
2N3456	NJD	B15	50V	4MA	150C	0.3WF	4.8MXV	0.2/1MA	0.3/0.9MO		5P	1P5	ALG	SIU	BF800	2N3686A	0
2N3457	NJD	B15	50V	4MA	150C	0.3WF	2.3MXV	2.5MXMA	0.15/.6MO		5P	1P5	ALG	SIU	BF800	2N3687A	0
2N3458	NJD	B6	50V	15MA	200C	0.3WF	7.8MXV	3/15MA	2.5/10MO		18P	5P	ALN	SIU	BF805	2N3684A	0
2N3459	NJD	B6	50V	15MA	200C	0.3WF	3.4MXV	0.8/4MA	1.5/6MO		18P	5P	ALN	SIU	BC264A	2N4868A	0

TYPE NO	CONS TRUC TION	CASE & LEAD	V DS MAX	I D MAX	T J MAX	P TOT MAX	VP OR VT	IDSS OR IDOM	G MO	R DS MAX	C ISS MAX	C RSS MAX	USE	SUPP LIER	EUR SUB	USA SUB	ISS
2N3460	NJD	B6	50V	15MA	200C	0.3WF	1.8MXV	0.2/1MA	0.8/4.5MO		18P	5P	ALN	SIU	BFW13	2N3686A	0
2N3465	NJD	B23	40V	5MA	200C	0.4WF	10MXV	5MXMA	0.4/1.2MO		15P		ALG	CRY	BF808	2N5457	0
2N3466	NJD	B47	40V	5MA	200C	0.4WF	10MXV	5MXMA	0.4/1.2MO		15P		ALG	CRY	BF808	2N5457	0
2N3573	PJD	B12	25V	10MA	200C	0.1WF	2MXV	0.1MXMA	0.1/0.3MO		6P		RLS	TIB		2N5020	0
2N3574	PJD	B12	25V	10MA	200C	0.1WF	2MXV	0.4MXMA	0.2/0.6MO		6P	2P	RLS	TDY		2N5020	0
2N3575	PJD	B12	25V	10MA	200C	0.2WF	2MXV	0.2/1MA	0.3/0.9MO		6P	2P	RLS	TDY		2N5020	0
2N3578	PJD	B7	20V	5MA	175C	0.3WF	1.5/4V	.9/4.5MA	1.2/3.5MO		65P		ALN	SIU	BF320A	2N3332	0
2N3608	PME	B79	25V	10MA	125C	.35WF	6MXV	7MXMA		400R			RLS	INB		3N174	0
2N3609	PME	B40	25V	5MA	125C	.35WF	6MXV	3.3MXMA		1K	2P		DUA	OBS		3N165	0
2N3610	PME	B53	20V		150C	0.1WF	7MXV	6MXMA		2K5		0P6	RLS	OBS		2N4352	0
2N3631	NMD	B7	20V	20MA	200C	0.3WF	6MXV	10MXMA	1.4/2.8MO		7P5		ALG	SIU	BFW96	2N3631	0
2N3631CHP	NMD	B73	20V	10MA	200C		6MXV	2/10MA	1.4/2.8MO	550R	7P5	1P6	ALG	SIU			0
2N3684	NJD	B15	50V	7.5MA	200C	0.3WF	2/5V	2/7.5MA	2/3MO	600R	4P	1P2	ALN	INB	BF805	2N3684A	0
2N3684/D	NJD	B73	50V	7.5MA	200C	0.3WF	2/5V	2/7.5MA	2/3MO	600R	4P	1P2	ALN	INB	BF805	2N3684A	0
2N3684/W	NJD	B74	50V	7.5MA	200C	0.3WF	2/5V	2/7.5MA	2/3MO	600R	4P	1P2	ALN	INB	BF805	2N3684A	0
2N3684A	NJD	B15	50V	7.5MA	200C	0.3WF	5MXV	2/7.5MA	2/3MO	600R	4P	1P2	ALN	INB	BF805	2N3684A	0
2N3684CHP	NJD	B73	50V	8MA	200C		2/5V	2.5/7MA	2/3MO		4P	1P2	ALG	SIU			0
2N3684T092	NJD	B63	50V	7.5MA	125C	0.3WF	2/5V	2/7.5MA	2/3MO	600R	4P	1P2	ALN	INB	BF805	2N3684A	0
2N3685	NJD	B15	50V	3MA	200C	0.3WF	1/3.5V	1/3MA	1.5/2.5MO	800R	4P	1P2	ALN	INB	BF808	2N3685A	0
2N3685/D	NJD	B73	50V	3MA	200C	0.3WF	1/3.5V	1/3MA	1.5/2.5MO	800R	4P	1P2	ALN	INB	BF808	2N3685A	0
2N3685/W	NJD	B74	50V	3MA	200C	0.3WF	1/3.5V	1/3MA	1.5/2.5MO	800R	4P	1P2	ALN	INB	BF808	2N3685A	0
2N3685A	NJD	B15	50V	3MA	200C	.15WF	3.5MXV	3MXMA	1.5/2.5MO		4P	1P2	ALN	INB	BF808	2N4868A	0
2N3685CHP	NJD	B73	50V	3MA	200C		1/3.5V	1/3MA	1.5/2.5MO		4P	1P2	ALG	SIU			0
2N3685T092	NJD	B63	50V	3MA	125C	0.3WF	1/3.5V	1/3MA	1.5/2.5MO	800R	4P	1P2	ALN	INB	BF808	2N3685A	0
2N3686	NJD	B15	50V	1.2MA	200C	0.3WF	0.6/2V	.4/1.2MA	1/2MO	1K2	4P	1P2	ALN	INB	BF800	2N3686A	0

TYPE NO	CONS TRUC TION	CASE & LEAD	V DS MAX	I D MAX	T J MAX	P TOT MAX	VP OR VT	IDSS OR IDOM	G MO	R DS MAX	C ISS MAX	C RSS MAX	USE	SUPP LIER	EUR SUB	USA SUB	ISS
2N3686/D	NJD	B73	50V	1.2MA	200C	0.3WF	0.6/2V	.4/1.2MA	1/2MO	1K2	4P	1P2	ALN	INB	BF800	2N3686A	0
2N3686/W	NJD	B74	50V	1.2MA	200C	0.3WF	0.6/2V	.4/1.2MA	1/2MO	1K2	4P	1P2	ALN	INB	BF800	2N3686A	0
2N3686A	NJD	B15	50V	2MA		0.3WF	2MXV	.6/1.2MA	1/2MO		4P	1P2	ALN	INB	BF800	2N4867A	0
2N3686CHP	NJD	B73	50V	2MA	200C		0.6/2V	.4/1.2MA	1/2MO	1K2	4P	1P2	ALG	SIU			0
2N3686T092	NJD	B63	50V	1.2MA	125C	0.3WF	0.6/2V	.4/1.2MA	1/2MO	1K2	4P	1P2	ALN	INB	BF800	2N3686A	0
2N3687	NJD	B15	50V	0.5MA	200C	0.3WF	0.3/1.2V	.1/0.5MA	0.5/1.5MO	2K4	4P	1P2	ALN	INB	BF800	2N3687A	0
2N3687/D	NJD	B73	50V	0.5MA	200C	0.3WF	0.3/1.2V	.1/0.5MA	0.5/1.5MO	2K4	4P	1P2	ALN	INB	BF800	2N3687A	0
2N3687/W	NJD	B74	50V	0.5MA	200C	0.3WF	0.3/1.2V	.1/0.5MA	0.3/1.2MO	2K4	4P	1P2	ALN	INB	BF800	2N3687A	0
2N3687A	NJD	B15	50V	1MA	200	0.3WF	1.2MXV	0.1/.5MA	0.5/1.5MO		4P	1P2	ALN	INB	BF800	2N4867A	0
2N3687CHP	NJD	B73	50V	1MA	200C		0.3/1.2V	0.1/.5MA	0.5/1.5MO		4P	1P2	ALG	SIU			0
2N3687T092	NJD	B63	50V	0.5MA	125C	0.3WF	0.3/1.2V	.1/0.5MA	0.5/1.5MO	2K4	4P	1P2	ALN	INB	BF800	2N3687A	0
2N3695	PJD	B12	30V	5MA	200C	0.1WF	4.5MXV	3.7MXMA	1/1.7MO		5P		ALN	OBS	BF320A	2N3329	0
2N3696	PJD	B12	20V	5MA	200C	0.3WF	3.2MXV	1.5MXMA	.75/1.2MO		5P		ALN	OBS	BF320A	2N5265	0
2N3697	PJD	B12	20V	5MA	200C	0.3WF	1.8MXV	0.6MXMA	0.5/1MO		5P		ALN	OBS		2N5265	0
2N3698	PJD	B12	30V	1MA	200C	0.3WF	1.2MXV	0.25MXMA	.25/.75MO		5P		ALN	OBS		2N3574	0
2N3796	NMD	B65	25V	20MA	200C	0.3WF	4MXV	0.3/3MA	0.9/1.8MO		7P	0P8	ALG	MOB		2N3631	0
2N3797	NMD	B65	20V	20MA	200C	0.3WF	7MXV	2/6MA	1.5/2.3MO		8P	0P8	ALG	MOB		2N3631	0
2N3819	NJD	B1	25V	20MA	150C	.36WF	8MXV	2/20MA	2/6.5MO		8P	4P	ALG	TIB	BF244	2N5459	0
2N3820	PJD	B1	20V	15MA	150C	.36WF	8MXV	0.3/15MA	0.8/5MO		32P	16P	ALG	TIB	BF320	2N3909	0
2N3821	NJD	B15	50V	2.5MA	200C	0.3WF	4MXV	.5/2.5MA	1.5/4.5MO		6P	3P	ALG	INB	BF347	2N4302	0
2N3821/D	NJD	B73	50V	2.5MA	200C	0.3WF	4MXV	.5/2.5MA	1.5/4.5MO		6P	3P	ALG	INB	BF347	2N4302	0
2N3821/W	NJD	B74	50V	2.5MA	200C	0.3WF	4MXV	.5/2.5MA	1.5/4.5MO		6P	3P	ALG	INB	BF347	2N4302	0
2N3821CHP	NJD	B73	50V	3MA	200C		0.5/2V	.5/2.5MA	1.5/4.5MO		6P	3P	ALG	SIU			0
2N3821JAN	NJD	B15	50V	5MA	200C	0.3WF	4MXV	2.5MXMA	1.5/4.5MO		6P		ALG	TIB	BF347	2N5457	0
2N3821T092	NJD	B63	50V	2.5MA	125C	0.3WF	4MXV	.5/2.5MA	1.5/4.5MO		6P	3P	ALG	INB	BF347	2N5457	0

TYPE NO	CONS TRUC TION	CASE & LEAD	V DS MAX	I D MAX	T J MAX	P TOT MAX	VP OR VT	IDSS OR IDOM	G MO	R DS MAX	C ISS MAX	C RSS MAX	USE	SUPP LIER	EUR SUB	USA SUB	ISS
2N3822	NJD	B15	50V	10MA	200C	0.3WF	6MXV	2/10MA	3/6.5MO		6P	3P	ALG	INB	BFW61	2N4303	0
2N3822/D	NJD	B73	50V	10MA	200C	0.3WF	6MXV	2/10MA	3/6.5MO		6P	3P	ALG	INB	BFW61	2N4303	0
2N3822/W	NJD	B74	50V	10MA	200C	0.3WF	6MXV	2/10MA	3/6.5MO		6P	3P	ALG	INB	BFW61	2N4303	0
2N3822CHP	NJD	B73	50V	10MA	200C		1/4V	2/10MA	3/6.5MO		6P	3P	ALG	SIU			0
2N3822JAN	NJD	B15	50V	10MA	200C	0.3WF	6MXV	10MXMA	3/6.5MO		6P		ALG	INB	BFW61	2N5459	0
2N3822T092	NJD	B63	50V	10MA	125C	0.3WF	6MXV	2/10MA	3/6.5MO		6P	3P	ALG	INB	BF244	2N5459	0
2N3823	NJD	B15	30V	20MA	175C	0.3WF	8MXV	4/20MA	3.5/6.5MO		6P	2P	FVG	SIU	BFW61	2N4224	0
2N3823CHP	NJD	B73	30V	20MA	200C		8MXV	4/20MA	3.5/6.5MO		6P	2P	RLG	SIU			0
2N3823JAN	NJD	B15	50V	10MA	200C	0.3WF	8MXV	20MXMA	3.5/6.5MO		6P		ALG	TIB	BFW61	2N3819	0
2N3824	NJD	B15	50V	20MA	200	0.3WF	8MXV	12/24MA	4.5MNMO	250R	6P	3P	RLS	SIU	BF244	2N5555	0
2N3824CHP	NJD	B73	50V	20MA	200C		8MXV	12/24MA	4.5MNMO	250R	6P	3P	RLS	SIU			0
2N3882	PME	B80	30V	1MA	150C	0.2WF	3MXV	1MXMA	1TPMO				RLS	OBS	BSW95	2N4352	0
2N3909	PJD	B12	20V	15MA	175C	0.3WF	8MXV	0.3/15MA	1/5MO		32P	16P	ALG	SIU	BF320	2N3331	0
2N3909A	PJD	B12	20V	15MA	200C	0.3WF	8MXV	1/15MA	2.2/5MO		9P	3P	ALG	MOB	BF320	2N5270	0
2N3921	NJD	B20	50V	10MA	200C	0.3WF	3MXV	1/10MA	1.5/7.5MO		18P	6P	DUA	SIU	BFQ10	2N5545	0
2N3921X2	NJD	B20	50V	10MA	200C	0.3WF	3MXV	1/10MA	1.5/7.5MO		18P	6P	DUA	SIU	BFQ10	2N5545	0
2N3922	NJD	B20	50V	10MA	200C	0.3WF	3MXV	1/10MA	1.5/7.5MO		18P	6P	DUA	SIU	BFQ10	2N5546	0
2N3922X2	NJD	B20	50V	10MA	200C	0.3WF	3MXV	1/10MA	1.5/7.5MO		18P	6P	DUA	SIU	BFQ10	2N5546	0
2N3934	NJD	B51	40V	1.3MA	175C	0.2WF	3MXV	.3/1.3MA	0.3/0.9MO		7P	2P2	DUA	TDY	BFQ10	2N4082	0
2N3935	NJD	B51	40V	1.3MA	175C	0.2WF	3MXV	.3/1.3MA	0.3/0.9MO		7P	2P2	DUA	TDY	BFQ10	2N4082	0
2N3935X2	NJD	B51	40V	1.3MA	175C	0.2WF	3MXV	.3/1.3MA	0.3/0.9MO		7P	2P2	DUA	TDY	BFQ10	2N4082	0
2N3954	NJD	B51	50V	5MA	125C	0.5WF	1/4.5V	0.5/5MA	1/3MO		4P	1P2	DUA	INB	BFQ10	2N5452	0
2N3954/D	NJD	B73	50V	5MA	125C		1/4.5V	0.5/5MA	1/3MO		4P	1P2	DUA	INB	BFQ10	2N5452	0
2N3954/W	NJD	B74	50V	5MA	125C		1/4.5V	0.5/5MA	1/3MO		4P	1P2	DUA	INB	BFQ10	2N5452	0
2N3954A	NJD	B51	50V	5MA	125C	0.5WF	1/4.5V	0.5/5MA	1/3MO		4P	1P2	DUA	INB	BFQ10	2N5452	0

TYPE NO	CONS TRUC TION	CASE & LEAD	V DS MAX	I D MAX	T J MAX	P TOT MAX	VP OR VT	IDSS OR IDOM	G MO	R DS MAX	C ISS MAX	C RSS MAX	USE	SUPP LIER	EUR SUB	USA SUB	ISS
2N3954A/D	NJD	B73	50V	5MA	125C		1/4.5V	0.5/5MA	1/3MO		4P	1P2	DUA	INB	BFQ10	2N5452	0
2N3954A/W	NJD	B51	50V	5MA	125C		1/4.5V	0.5/5MA	1/3MO		4P	1P2	DUA	INB	BFQ10	2N5452	0
2N3954AX2	NJD	B51	50V	5MA	125C	0.5WF	1/4.5V	0.5/5MA	1/3MO		4P	1P2	DUA	INB	BFQ10	2N5452	0
2N3954X2	NJD	B51	50V	5MA	125C	0.5WF	1/4.5V	0.5/5MA	1/3MO		4P	1P2	DUA	INB	BFQ10	2N5452	0
2N3955	NJD	B51	50V	5MA	125C	0.5WF	1/4.5V	0.5/5MA	1/3MO		4P	1P2	DUA	INB	BFQ10	2N3954	0
2N3955/D	NJD	B73	50V	5MA	125C		1/4.5V	0.5/5MA	1/3MO		4P	1P2	DUA	INB	BFQ10	2N3954	0
2N3955/W	NJD	B74	50V	5MA	125C		1/4.5V	0.5/5MA	1/3MO		4P	1P2	DUA	INB	BFQ10	2N3954	0
2N3955A	NJD	B51	50V	5MA	125C	0.5WF	1/4.5V	5MA	1/3MO		4P	1P2	DUA	INB	BFQ10	2N3954	0
2N3955A/D	NJD	B73	50V	5MA	125C		1/4.5V	5MA	1/3MO		4P	1P2	DUA	INB	BFQ10	2N3954	0
2N3955A/W	NJD	B74	50V	5MA	125C		1/4.5V	5MA	1/3MO		4P	1P2	DUA	INB	BFQ10	2N3954	0
2N3955AX2	NJD	B51	50V	5MA	125C	0.5WF	1/4.5V	5MA	1/3MO		4P	1P2	DUA	INB	BFQ10	2N3954	0
2N3955X2	NJD	B51	50V	5MA	125C	0.5WF	1/4.5V	0.5/5MA	1/3MO		4P	1P2	DUA	INB	BFQ10	2N3954	0
2N3956	NJD	B51	50V	5MA	125C	0.5WF	1/4.5V	0.5/5MA	1/3MO		4P	1P2	DUA	INB	BFQ10	2N3954	0
2N3956/D	NJD	B73	50V	5MA	125C		1/4.5V	0.5/5MA	1/3MO		4P	1P2	DUA	INB	BFQ10	2N3954	0
2N3956/W	NJD	B74	50V	5MA	125C		1/4.5V	0.5/5MA	1/3MO		4P	1P2	DUA	INB	BFQ10	2N3954	0
2N3956X2	NJD	B51	50V	5MA	125C	0.5WF	1/4.5V	0.5/5MA	1/3MO		4P	1P2	DUA	INB	BFQ10	2N3954	0
2N3957	NJD	B51	50V	5MA	125C	0.5WF	1/4.5V	0.5/5MA	1/3MO		4P	1P2	DUA	INB	BFQ10	2N3954	0
2N3957/D	NJD	B73	50V	5MA	125C		1/4.5V	0.5/5MA	1/3MO		4P	1P2	DUA	INB	BFQ10	2N3954	0
2N3957/W	NJD	B74	50V	5MA	125C		1/4.5V	0.5/5MA	1/3MO		4P	1P2	DUA	INB	BFQ10	2N3954	0
2N3957X2	NJD	B51	50V	5MA	125C	0.5WF	1/4.5V	0.5/5MA	1/3MO		4P	1P2	DUA	INB	BFQ10	2N3954	0
2N3958	NJD	B51	50V	5MA	125C	0.5WF	1/4.5V	0.5/5MA	1/3MO		4P	1P2	DUA	INB	BFQ10	2N3954	0
2N3958/D	NJD	B73	50V	5MA	125C		1/4.5V	0.5/5MA	1/3MO		4P	1P2	DUA	INB	BFQ10	2N3954	0
2N3958/W	NJD	B74	50V	5MA	125C		1/4.5V	0.5/5MA	1/3MO		4P	1P2	DUA	INB	BFQ10	2N3954	0
2N3958X2	NJD	B51	50V	5MA	125C	0.5WF	1/4.5V	0.5/5MA	1/3MO		4P	1P2	DUA	INB	BFQ10	2N3954	0
2N3966	NJD	B15	30V	20MA	175C	0.3WF	4/6V	2MNMA		220R	6P	1P5	RLS	INB		2N3824	0

TYPE NO	CONS TRUC TION	CASE & LEAD	V DS MAX	I D MAX	T J MAX	P TOT MAX	VP OR VT	IDSS OR IDOM	G MO	R DS MAX	C ISS MAX	C RSS MAX	USE	SUPP LIER	EUR SUB	USA SUB	ISS
2N3967	NJD	B15	30V	10MA	175C	0.3WF	2/5V	2.5/10MA	2.5MNMO	400R	5P	1P3	RLS	TDY	BFW61	2N3822	0
2N3967A	NJD	B15	30V	10MA	175C	0.3WF	2/5V	2.5/10MA	2.5MNMO	400R	5P	1P3	RLS	TDY	BFW61	2N3822	0
2N3968	NJD	B15	30V	5MA	175C	0.3WF	3MXV	1/5MA	2MNMO	700R	5P	1P3	ALG	TDY	BFW12	2N4303	0
2N3968A	NJD	B15	30V	5MA	175C	0.3WF	3MXV	1/5MA	2MNMO	700R	5P	1P3	ALG	TDY	BFW12	2N3821	0
2N3969	NJD	B15	30V	2MA	175C	0.3WF	1.7MXV	0.4/2MA	1.3MNMO		5P	1P3	ALG	TDY	BF347	2N3821	0
2N3969A	NJD	B15	30V	2MA	175C	0.3WF	1.7MXV	0.4/2MA	1.3MNMO		5P	1P3	ALG	TDY	BF347	2N3821	0
2N3970	NJD	B6	40V	150MA	200C	1.8WC	4/10V	50/150MA		30R	25P	6P	RLS	SIU	BSV78	2N4391	0
2N3971	NJD	B6	40V	150MA	200C	1.8WC	2/5V	25/75MA		60R	25P	6P	RLS	SIU	BSV79	2N4392	0
2N3972	NJD	B6	40V	150MA	200C	1.8WC	0.5/3V	5/30MA		100R	25P	6P	RLS	SIU	BSV80	2N4393	0
2N3993	PJD	B15	25V	50MA	200C	0.3WF	4/2.5V	10MNMA	6/12MO	150R	16P	4P5	RLS	INB		2N3993A	0
2N3993A	PJD	B12	25V	10MA	200C	0.3WF		10MXMA	7/12MO	150R	12P		RLS	INB	BFT11	2N3382	0
2N3994	PJD	B15	25V	50MA	200C	0.3WF	1/5.5V	2MNMA	4/10MO	300R	16P	4P5	RLS	INB		2N3994A	0
2N3994A	PJD	B12	25V	50MA	200C	0.3WF	5.5MXV	2MNMA	5/10MO	300R	12P	3P5	RLS	MOB	BFT11	2N3934A	0
2N4038	NME	B56	25V	20MA	175C	.12WF		0.1MXMA	1.5/2.5MO	20K	2P5		RLS	TRW		2N4351	0
2N4039	NMD	B56	25V	20MA	175C	.12WF		1.5MXMA	1.2/2.5MO	20K	2P5		ALG	TRW	BFX63	2N3631	0
2N4065	PME	B53	30V	20MA	175C	.35WF	6MXV		0.4MNMO	1K5	4P5		RLS	GIB	BSX83	3N174	0
2N4066	PME	B38	30V	200MA	200C	0.6WF	3/6V	10/50MA	1.5MNMO	500R	7P	1P5	DUA	MOB		3N165	0
2N4067	PME	B38	30V	200MA	200C	0.6WF	3/6V	10/50MA	2.5MNMO	250R	7P	1P5	DUA	MOB		3N166	0
2N4082	NJD	B51	50V	2MA	200C	0.3WF	3MXV	.3/1.3MA	0.3MNMO		7P	2P2	DUA	TDY	BFQ10	2N3954	0
2N4082X2	NJD	B51	50V	2MA	200C	0.3WF	3MXV	.3/1.3MA	0.3MNMO		7P	2P2	DUA	TDY	BFQ10	2N3954	0
2N4083	NJD	B51		2MA	200C	0.3WF	3MXV	.3/1.3MA	0.3MNMO		7P	2P2	DUA	TDY		2N4082	0
2N4083X2	NJD	B51		2MA	200C	0.3WF	3MXV	.3/1.3MA	0.3MNMO		7P	2P2	DUA	TDY		2N4082	0
2N4084	NJD	B20	50V	10MA	200C	0.3WF	3MXV	1/10MA	1.5/7.5MO		18P	6P	DUA	SIU	BFQ10	2N5545	0
2N4084X2	NJD	B20	50V	10MA	200C	0.3WF	3MXV	1/10MA	1.5/7.5MO		18P	6P	DUA	SIU	BFQ10	2N5545	0
2N4085	NJD	B20	50V	10MA	200C	0.3WF	3MXV	1/10MA	1.5/7.5MO		18P	6P	DUA	SIU	BFQ10	2N5546	0

TYPE NO	CONS TRUC TION	CASE & LEAD	V DS MAX	I D MAX	T J MAX	P TOT MAX	VP OR VT	IDSS OR IDOM	G MO	R DS MAX	C ISS MAX	C RSS MAX	USE	SUPP LIER	EUR SUB	USA SUB	ISS
2N4085X2	NJD	B20	50V	10MA	200C	0.3WF	3MXV	1/10MA	1.5/7.5MO		18P	6P	DUA	SIU	BFQ10	2N5546	0
2N4088	PJD	B15	30V	25MA	200C	0.3WF	8MXV	15MXMA	1/1.6MO		10P		ALG	OBS	BF320B	2N2386	0
2N4089	PJD	B15	30V	10MA	200C	0.3WF	5MXV	8MXMA	0.8/1.3MO		10P		ALG	TDY	BF320B	2N3330	0
2N4090	PJD	B15	30V	10MA	200C	0.3WF	3MXV	2.5MXMA	0.5/.9MO		10P		ALG	TDY	BF320A	2N3329	0
2N4091	NJD	B6	40V	150MA	175C	.36WF	5/10V	30MNMA		30R	16P	5P	RLS	INB	BSV78	2N4391	0
2N4091/D	NJD	B73	40V	150MA	175C		5/10V	30MNMA		30R	16P	5P	RLS	INB	BSV78	2N4391	0
2N4091/W	NJD	B74	40V	150MA	175C		5/10V	30MNMA		30R	16P	5P	RLS	INB	BSV78	2N4391	0
2N4091A	NJD	B6	50V	100MA	200C	1.8WC	2/7V	30MNMA		30R	16P	5P	RMS	OBS	BSV78	2N4856	0
2N4091CHP	NJD	B73	40V	150MA	200C		5/10V	30MNMA		30R	16P	5P	RLS	SIU			0
2N4091JAN	NJD	B6	40V	150MA	175C	.36WF	5/10V	30MNMA		30R	16P	5P	RLS	INB	BSV78	2N4391	0
2N4091T092	NJD	B63	40V	150MA	175C	.36WF	5/10V	30MNMA		30R	16P	5P	RLS	INB	BSV78	2N4391	0
2N4092	NJD	B6	40V	150MA	175C	.36WF	2/7V	15MNMA		50R	16P	5P	RLS·	INB	BSV79	2N4392	0
2N4092/D	NJD	B73	40V	150MA	175C		2/7V	15MNMA		50R	16P	5P	RLS	INB	BSV79	2N4392	0
2N4092/W	NJD	B74	40V	150MA	175C		2/7V	15MNMA		50R	16P	5P	RLS	INB	BSV79	2N4392	0
2N4092A	NJD	B6	50V	100MA	200C	1.8WC	5/10V	15MNMA		50R	16P	5P	RMS	OBS	BSV79	2N4857	0
2N4092CHP	NJD	B73	40V	A	200C		2/7V	15MNMA		50R	16P	5P	RLS	SIU			0
2N4092JAN	NJD	B6	40V	150MA	175C	.36WF	2/7V	15MNMA		50R	16P	5P	RLS	INB	BSV79	2N	0
2N4092T092	NJD	B63	40V	150MA	175C	.36WF	2/7V	15MNMA		50R	16P	5P	RLS	INB	BSV79	2N5555	0
2N4093	NJD	B6	40V	150MA	175C	.36WF	1/5V	8MNMA		80R	16P	5P	RLS	INB	BSV80	2N4858	0
2N4093/D	NJD	B73	40V	150MA	175C		1/5V	8MNMA		80R	16P	5P	RLS	INB	BSV80	2N4858	0
2N4093/W	NJD	B74	40V	150MA	175C		1/5V	8MNMA		80R	16P	5P	RLS	INB	BSV80	2N4858	0
2N4093A	NJD	B6	50V	100MA	200C	1.8WC	1/5V	8MNMA		80R	16P	5P	RMS	OBS	BSV80	2N4858	0
2N4093CHP	NJD	B73	40V	150MA	175C		1/5V	8MNMA		80R	16P	5P	RLS	SIU			0
2N4093JAN	NJD	B6	40V	150MA	175C	.36WF	1/5V	8MNMA		80R	16P	5P	RLS	INB	BSV80	2N3824	0
2N4093T092		B63	40V	150MA	175C	.36WF	1/5V	8MNMA		80R	16P	5P	RLS	INB	BSV80	2N5555	0

TYPE NO	CONS TRUC TION	CASE & LEAD	V DS MAX	I D MAX	T J MAX	P TOT MAX	VP OR VT	IDSS OR IDOM	G MO	R DS MAX	C ISS MAX	C RSS MAX	USE	SUPP LIER	EUR SUB	USA SUB	ISS
2N4094	NJD	B6	40V	150MA	200C	1.8WC		75MNMA		20R	32P		RMS	OBS	BSV78	2N5434	0
2N4095	NJD	B6	40V	150MA	200C	1.8WC		20MNMA		40R	32P		RMS	OBS	BSV79	2N4857	0
2N4117	NJD	B15	40V	1MA	200C	0.3WF	0.6/1.8V	.09MXMA	.07/.21MO		3P	1P5	ALG	INB	BFW13	2N3821	0
2N4117/D	NJD	B73	40V	1MA	200C		0.6/1.8V	.09MXMA	.07/.21MO		3P	1P5	ALG	INB	BFW13	2N3821	0
2N4117/W	NJD	B74	40V	1MA	200C		0.6/1.8V	.09MXMA	.07/.21MO		3P	1P5	ALG	INB	BFW13	2N3821	0
2N4117A	NJD	B15	40V	1MA	200C	0.3WF	0.6/1.8V	.09MXMA	.07/.21MO		3P	1P5	ALG	INB	BFW13	2N3821	0
2N4117A/D	NJD	B73	40V	1MA	200C		0.6/1.8V	.09MXMA	.07/.21MO		3P	1P5	ALG	INB	BFW13	2N3821	0
2N4117A/W	NJD	B74	40V	1MA	200C		0.6/1.8V	.09MXMA	.07/.21MO		3P	1P5	ALG	INB	BFW13	2N3821	0
2N4117AT92	NJD	B63	40V	1MA	125C	0.3WF	0.6/1.8V	.09MXMA	.07/.21MO		3P	1P5	ALG	INB	BF347	2N5457	0
2N4117CHP	NJD	B73	40V	1MA	200C		0.6/1.8V	.03/.1MA	.07/.21MO		3P	1P5	ALN	SIU			0
2N4117T092	NJD	B63	40V	1MA	200C		0.6/1.8V	.09MXMA	.07/.21MO		3P	1P5	ALG	INB	BF347	2N5457	0
2N4118	NJD	B15	40V	1MA	200C	0.3WF	1/3V	.24MXMA	.08/.25MO		3P	1P5	ALG	INB	BFW13	2N3821	0
2N4118/D	NJD	B73	40V	1MA	200C		1/3V	.24MXMA	.08/.25MO		3P	1P5	ALG	INB	BFW13	2N3821	0
2N4118/W	NJD	B74	40V	1MA	200C		1/3V	.24MXMA	.08/.25MO		3P	1P5	ALG	INB	BFW13	2N3821	0
2N4118A	NJD	B15	40V	1MA	200C	0.3WF	1/3V	.24MXMA	.08/.25MO		3P	1P5	ALG	INB	BFW13	2N3821	0
2N4118A/D	NJD	B73	40V	1MA	200C		1/3V	.24MXMA	.08/.25MO		3P	1P5	ALG	INB	BFW13	2N3821	0
2N4118A/W	NJD	B74	40V	1MA	200C		1/3V	.24MXMA	.08/.25MO		3P	1P5	ALG	INB	BFW13	2N3821	0
2N4118AT92	NJD	B63	40V	1MA	125C	0.3WF	1/3V	.24MXMA	.08/.25MO		3P	1P5	ALG	INB	BF347	2N5457	0
2N4118CHP	NJD	B73	40V	1MA	200C		1/3V	0.24MXMA	.08/.25MO		3P	1P5	ALN	SIU			0
2N4118T092	NJD	B63	40V	1MA	125C	0.3WF	1/3V	.24MXMA	.08/.25MO		3P	1P5	ALG	INB	BF347	2N5457	0
2N4119	NJD	B15	40V	1MA	200C	0.3WF	2/6V	.2/0.6MA	.1/.33MO		3P	1P5	ALG	INB	BFW13	2N3821	0
2N4119/D	NJD	B73	40V	1MA	200C		2/6V	.2/0.6MA	.1/.33MO		3P	1P5	ALG	INB	BFW13	2N3821	0
2N4119/W	NJD	B74	40V	1MA	200C		2/6V	.2/0.6MA	.1/.33MO		3P	1P5	ALG	INB	BFW13	2N3821	0
2N4119A	NJD	B15	40V	1MA	200C	0.3WF	2/6V	.2/0.6MA	.1/.33MO		3P	1P5	ALG	INB	BFW13	2N3821	0
2N4119A/D	NJD	B73	40V	1MA	200C		2/6V	.2/0.6MA	.1/.33MO		3P	1P5	ALG	INB	BFW13	2N3821	0

TYPE NO	CONS TRUC TION	CASE & LEAD	V DS MAX	I D MAX	T J MAX	P TOT MAX	VP OR VT	IDSS OR IDOM	G MO	R DS MAX	C ISS MAX	C RSS MAX	USE	SUPP LIER	EUR SUB	USA SUB	ISS
2N4119A/W	NJD	B74	40V	1MA	200C		2/6V	.2/0.6MA	.1/.33MO		3P	1P5	ALG	INB	BFW13	2N3821	0
2N4119AT92	NJD	B63	40V	1MA	125C	0.3WF	2/6V	.2/0.6MA	.1/.33MO		3P	1P5	ALG	INB	BF347	2N5457	0
2N4119CHP	NJD	B73	40V	1MA	200C		2/6V	0.2/.6MA	0.1/.33MO		3P	1P5	ALN	SIU			0
2N4119T092	NJD	B63	40V	1MA	125C	0.3WF	2/6V	.2/0.6MA	.1/.33MO		3P	1P5	ALG	INB	BFW13	2N3821	0
2N4120	PME	B53	30V	20MA	175C	.35WF	6MXV	12MXMA	0.7MNMO	1K	4P5		RLS	GIB	BSX83	3N174	0
2N4139	NJD	B6	50V	11MA	200C	0.3WF	2/8V	8/11MA	3.5/7.5MO		18P	5P	RLS	TDY	BF805	2N5396	0
2N4220	NJD	B15	30V	15MA	200C	0.3WF	4MXV	0.5/3MA	1/4MO		6P	2P	FVG	INB	BFW61	2N5104	0
2N4220/D	NJD	B73	30V	15MA	200C		4MXV	0.5/3MA	1/4MO		6P	2P	FVG	INB	BFW61	2N5484	0
2N4220/W	NJD	B74	30V	15MA	200C		4MXV	0.5/3MA	1/4MO		6P	2P	FVG	INB	BFW61	2N5484	0
2N4220A	NJD	B14	30V	15MA	175C	0.3WF	4MXV	0.5/3MA	1/4MO		6P	2P	ALN	SIU	BF347	2N4302	0
2N4220T092	NJD	B63	30V	15MA	125C	0.3WF	4MXV	0.5/3MA	1/4MO		6P	2P	FVG	INB	BFW61	2N5484	0
2N4221	NJD	B15	30V	15MA	200C	0.3WF	6MXV	2/6MA	2/5MO		6P	2P	FVG	INB	BFW61	2N5104	0
2N4221/D	NJD	B73	30V	15MA	200C		6MXV	2/6MA	2/5MO		6P	2P	FVG	INB	BFW61	2N5104	0
2N4221/W	NJD	B74	30V	15MA	200C		6MXV	2/6MA	2/5MO		6P	2P	FVG	INB	BFW61	2N5104	0
2N4221A	NJD	B14	30V	15MA	175C	0.3WF	6MXV	2/6MA	2/5MO		6P	2P	ALN	SIU	BC264B	2N3684A	0
2N4221T092	NJD	B63	30V	15MA	125C	0.3WF	6MXV	2/6MA	2/5MO		6P	2P	FVG	INB	BFW61	2N5484	0
2N4222	NJD	B15	30V	15MA	200C	0.3WF	8MXV	5/15MA2	.5/6MO		6P	2P	FVG	INB	BFW61	2N5105	0
2N4222/D	NJD	B73	30V	15MA	200C		8MXV	5/15MA	2.5/6MO		6P	2P	FVG	INB	BFW61	2N5245	0
2N4222/W	NJD	B74	30V	15MA	200C		8MXV	5/15MA	2.5/6MO		6P	2P	FVG	INB	BFW61	2N5245	0
2N4222A	NJD	B14	30V	15MA	175C	0.3WF	8MXV	5/15MA	2.5/6MO		6P	2P	ALN	SIU	BF805	2N5459	0
2N4222T092	NJD	B63	30V	15MA	125C	0.3WF	8MXV	5/15MA	2.5/6MO		6P	2P	FVG	INB	BFW61	2N5245	0
2N4223	NJD	B15	30V2	0MA	200C	0.3WF	0.1/8V	3/18MA	3/7MO		6P	2P	FVG	INB	BFW61	2N4224	0
2N4224	NJD	B15	30V	20MA	200C	0.3WF	0.1/8V	2/20MA	2/7.5MO		6P	2P	FVG	INB	BFW61	2N3823	0
2N4267	PME	B52	30V	500MA	175C	0.4WF	6MXV	100MXMA	3MNMO	250R	15P		RLS	INB		3N161	0
2N4268	PME	B52	30V	500MA	175C	0.4WF	6MXV	100MXMA	5MNMO	125R	15P		RLS	INB		3N161	0

TYPE NO	CONS TRUC TION	CASE & LEAD	V DS MAX	I D MAX	T J MAX	P TOT MAX	VP OR VT	IDSS OR IDOM	G MO	R DS MAX	C ISS MAX	C RSS MAX	USE	SUPP LIER	EUR SUB	USA SUB	ISS
2N4302	NJD	B10	30V	5MA	125C	0.3WF	4MXV	0.5/5MA	1MNMO		6P	3P	ALG	TDY	BF347	2N3686A	0
2N4303	NJD	B10	30V	10MA	125C	0.3WF	6MXV	4/10MA	2MNMO		6P	3P	ALG	TDY	BC264	2N3684A	0
2N4304	NJD	B10	30V	15MA	125C	0.3WF	10MXV	0.5/15MA	1MNMO		6P	3P	ALG	TDY	BC264C	2N5459	0
2N4338	NJD	B6	50V	9MA	200C	0.3WF	0.3/1V	.2/0.6MA	0.6/1.8MO	2K5	7P	3P	RLS	INB		2N3966	0
2N4338/D	NJD	B73	50V	9MA	200C		0.3/1V	.2/0.6MA	0.6/1.8MO	2K5	7P	3P	RLS	INB		2N3966	0
2N4338/W	NJD	B74	50V	9MA	200C		0.3/1V	.2/0.6MA	0.6/1.8MO	2K5	7P	3P	RLS	INB		2N3966	0
2N4338CHP	NJD	B73	50V	1MA	200C		0.3/1V	0.2/.6MA	0.6/1.8MO	2K5	7P	3P	ALG	SIU			0
2N4338T092	NJD	B63	50V	9MA	125C	0.3WF	0.3/1V	.2/0.6MA	0.6/1.8MO	2K5	7P	3P	RLS	INB		2N3966	0
2N4339	NJD	B6	50V	9MA	200C	0.3WF	0.6/1.8V	.5/1.5MA	0.8/2.4MO	1K7	7P	3P	RLS	INB		2N3966	0
2N4339/D	NJD	B73	50V	9MA	200C		0.6/1.8V	.5/1.5MA	0.8/2.4MO	1K7	7P	3P	RLS	INB		2N3966	0
2N4339/W	NJD	B74	50V	9MA	200C		0.6/1.8V	.5/1.5MA	0.8/2.4MO	1K7	7P	3P	RLS	INB		2N3966	0
2N4339CHP	NJD	B73	50V	2MA	200C		0.6/1.8V	.5/1.5MA	0.8/2.4MO	1K7	7P	3P	ALG	SIU			0
2N4339T092	NJD	B63	50V	9MA	125C	0.3WF	0.6/1.8V	.5/1.5MA	0.8/2.4MO	1K7	7P	3P	RLS	INB		2N3966	0
2N4340	NJD	B6	50V	9MA	200C	0.3WF	1/3V	1/3.6MA	1.3/3MO	1K5	7P	3P	RLS	INB		2N3966	0
2N4340/D	NJD	B73	50V	9MA	200C		1/3V	1/3.6MA	1.3/3MO	1K5	7P	3P	RLS	INB		2N3966	0
2N4340/W	NJD	B74	50V	9MA	200C		1/3V	1/3.6MA	1.3/3MO	1K5	7P	3P	RLS	INB		2N3966	0
2N4340CHP	NJD	B73	50V	4MA	200C		1/3V	3.6MXMA	1.3/3MO	1K5	7P	3P	ALG	SIU			0
2N4340T092	NJD	B63	50V	9MA	125C	0.3WF	1/3V	1/3.6MA	1.3/3MO	1K5	7P	3P	RLS	INB		2N3966	0
2N4341	NJD	B6	50V	9MA	200C	0.3WF	2/6V	3/9MA	2/4MO	800R	7P	3P	RLS	INB		2N4393	0
2N4341/D	NJD	B73	50V	9MA	200C		2/6V	3/9MA	2/4MO	800R	7P	3P	RLS	INB		2N3966	0
2N4341/W	NJD	B74	50V	9MA	200C		2/6V	3/9MA	2/4MO	800R	7P	3P	RLS	INB		2N3966	0
2N4341CHP	NJD	B73	50V	10MA	200C		2/6V	3/9MA	2/4MO	800R	7P	3P	ALG	SIU			0
2N4341T092	NJD	B63	50V	9MA	125C	0.3WF	2/6V	3/9MA	2/4MO	800R	7P	3P	RLS	INB		2N3966	0
2N4342	PJD	B10	25V	12MA	135C	.18WF	1/5.5V	4/12MA	2MNMO	700R	20P	5P	RLG	TDY	BF320C	2N2609	0
2N4343	PJD	B10	25V	30MA	135C	.18WF	2/10V	10/30MA	4MNMO	350R	20P	5P	RLG	TDY	BFT11	2N4343	0

TYPE NO	CONS TRUC TION	CASE & LEAD	V DS MAX	I D MAX	T J MAX	P TOT MAX	VP OR VT	IDSS OR IDOM	G MO	R DS MAX	C ISS MAX	C RSS MAX	USE	SUPP LIER	EUR SUB	USA SUB	ISS
2N4351	NME	B56	25V	100MA	150C	.37WF	1/5V	3MNMA	1MNMO	300R	5P1P	3	RLS	INB		2N4351	0
2N4351/D	NME	B73	25V	100MA	150C	.37WF	1/5V	3MNMA	1MNMO	300R	5P	1P3	RLS	INB		2N4351	0
2N4351/W	NME	B74	25V	100MA	150C	.37WF	1/5V	3MNMA	1MNMO	300R	5P	1P3	RLS	INB		2N4351	0
2N4352	PME	B12	25V	30MA	200C	0.3WF	1/5V	3MNMA	1MNMO	600R	5P	1P5	RLS	MOB		3N155	0
2N4353	PME	B53	30V	100MA	125C	0.2WF	5MXV		1/4MO	300R	12P		RLS	INB	BSW95	3N163	0
2N4360	PJD	B10	20V	30MA	180C	.25WF	0.7/10V	3/30MA	2/8MO	700R	20P	5P	RLG	TDY	BFT11	2N4343	0
2N4381	PJD	B7	25V	20MA	175C	0.3WF	5MXV	12MXMA	2/6MO	700R	20P		RLS	INB	BF320	2N3934A	0
2N4382	PJD	B7	25V	30MA	175C	0.3WF	9MXV	30MXMA	4/8MO	350R	20P		ALG	INB	BFT11	2N4343	0
2N4391	NJD	B6	40V	150MA	200C	0.3WF	4/10V	50/150MA		30R	14P	3P5	RMS	INB	BSV78	2N4856	0
2N4391/D	NJD	B73	40V	150MA	200C		4/10V	50/150MA		30R	14P	3P5	RLS	INB	BSV78	2N4856	0
2N4391/W	NJD	B74	40V	150MA	200C		4/10V	50/150MA		30R	14P	3P5	RLS	INB	BSV78	2N4856	0
2N4391CHP	NJD	B73	40V	150MA	200C		4/10V	50/150MA		30R	14P	3P5	RMS	SIU			0
2N4391T092	NJD	B63	40V	150MA	125C	0.3WF	4/10V	50/150MA		30R	14P	3P5	RLS	INB	BSV78	2N4856	0
2N4392	NJD	B6	40V	150MA	200C	0.3WF	2/5V	25/75MA		60R	14P	3P5	RLS	INB	BSV79	2N4091	0
2N4392/D	NJD	B73	40V	150MA	200C		2/5V	25/75MA		60R	14P	3P5	RLS	INB	BSV79	2N4091	0
2N4392/W	NJD	B74	40V	150MA	200C		2/5V	25/75MA		60R	14P	3P5	RLS	INB	BSV79	2N4091	0
2N4392CHP	NJD	B73	40V	150MA	200C		2/5V	25/75MA		60R	14P	3P5	RLS	SIU			0
2N4392T092	NJD	B63	40V	150MA	125C	0.3WF	2/5V	25/75MA		60R	14P	3P5	RLS	INB	BSV79	2N4091	0
2N4393	NJD	B6	40V	150MA	200C	0.3WF	0.5/3V	5/30MA		100R	14P	3P5	RLS	INB	BSV80	2N4093	0
2N4393/D	NJD	B73	40V	150MA	200C		0.5/3V	5/30MA		100R	14P	3P5	RLS	INB	BSV80	2N4093	0
2N4393/W	NJD	B74	40V	150MA	200C		0.5/3V	5/30MA		100R	14P	3P5	RLS	INB	BSV80	2N4093	0
2N4393CHP	NJD	B73	40V	30MA	200C		0.5/3V	5/30MA		100R	14P	3P5	RLS	SIU			0
2N4393T092	NJD	B63	40V	150MA	125C	0.3WF	0.5/3V	5/30MA		100R	14P	3P5	RLS	INB	BSV80	2N4093	0
2N4416	NJD	B15	30V	15MA	200C	0.3WF	2.5/6V	5/15MA4	.5/7.5MO		4P	0P8	TUG	INB	BF256	2N5245	0
2N4416/D	NJD	B73	30V	15MA	200C		2.5/6V	5/15MA	4.5/7.5MO		4P	0P8	TUG	INB	BF256	2N5245	0

TYPE NO	CONS TRUC TION	CASE & LEAD	V DS MAX	I D MAX	T J MAX	P TOT MAX	VP OR VT	IDSS OR IDOM	G MO	R DS MAX	C ISS MAX	C RSS MAX	USE	SUPP LIER	EUR SUB	USA SUB	ISS
2N4416/W	NJD	B74	30V	15MA	200C		2.5/6V	5/15MA	4.5/7.5MO		4P	OP8	TUG	INB	BF256	2N5245	0
2N4416A	NJD	B15	35V	15MA	200C	0.3WF	2.5/6V	5/15MA4	.5/7.5MO		4P0P	8	TUG	INB	BF256	2N5245	0
2N4416A/D	NJD	B73	35V	15MA	200C		2.5/6V	5/15MA	4.5/7.5MO		4P	OP8	TUG	INB	BF256	2N5245	0
2N4416A/W	NJD	B74	35V	15MA	200C		2.5/6V	5/15MA	4.5/7.5MO		4P	OP8	TUG	INB	BF256	2N5245	0
2N4416ACHP	NJD	B73	35V	15MA	200C		2/6V	5/15MA	4.5/7.5MO		4P	OP8	ULG	SIU			0
2N4416AJAN	NJD	B15	35V	15MA	200C	0.3WF	6MXV	5/15MA	4.5/7.5MO		4P	2P	TUG	INB	BF256	2N5245	0
2N4416AT92	NJD	B63	35V	15MA	125C	0.3WF	2.5/6V	5/15MA	4.5/7.5MO		4P	OP8	TUG	INB	BF256	2N5245	0
2N4416CHP	NJD	B73	30V	15MA	200C		6MXV	5/15MA	4.5/7.5MO		4P	OP8	ULG	SIU			0
2N4416T092	NJD	B63	30V	15MA	125C	0.3WF	2.5/6V	5/15MA	4.5/7.5MO		4P	OP8	TUG	INB	BF256	2N5245	0
2N4417	NJD	B69	30V	15MA	150C	.17WF	6MXV	5/15MA	4.5/7.5MO		3P5	OP8	TUG	SIU	BF256B	2N5245	0
2N4445	NJD	B47	25V	400MA	200C	0.4WF	2/10V	150MNMA		5R	25P		ALC	CRY		2N5432	0
2N4446	NJD	B47	25V	400MA	200C	0.4WF	2/10V	100MNMA		10R	25P		ALC	CRY		2N5432	0
2N4447	NJD	B47	25V	400MA	200C	0.4WF	2/10V	150MNMA		6R	25P		ALC	CRY		2N5432	0
2N4448	NJD	B47	25V	400MA	200C	0.4WF	2/10V	100MNMA		12R	25P		ALC	CRY		2N5432	0
2N4856	NJD	B6	40V	250MA	200C	1.8WC	4/10V	50MNMA		25R	18P	8P	RLS	INB	BSV78	2N4859	0
2N4856/D	NJD	B73	40V	250MA	200C		4/10V	50MNMA		25R	18P	8P	RLS	INB	BSV78	2N4859	0
2N4856/W	NJD	B74	40V	250MA	200C		4/10V	50MNMA		25R	18P	8P	RLS	INB	BSV78	2N4859	0
2N4856A	NJD	B6	40V	200MA	200C	1.8WC	4/10V	50MNMA		25R	10P	4P	RLS	TDY	BSV78	2N4856	0
2N4856CHP	NJD	B73	40V	250MA	200C		4/10V	50MNMA		25R	18P	8P	RMS	SIU			0
2N4856JAN	NJD	B6	40V	100MA	200C	.36WF	4/10V	50MNMA		25R	18P	8P	RMS	SIU	BSV78	2N4391	0
2N4856T092	NJD	B63	40V	250MA	125C	1.8WC	4/10V	50MNMA		25R	18P	8P	RLS	INB	BSV78	2N4859	0
2N4857	NJD	B6	40V	250MA	200C	1.8WC	2/6V	20/100MA		40R	18P	8P	RLS	INB	BSV79	2N4860	0
2N4857/D	NJD	B73	40V	250MA	200C		2/6V	20/100MA		40R	18P	8P	RLS	INB	BSV79	2N4860	0
2N4857/W	NJD	B74	40V	250MA	200C		2/6V	20/100MA		40R	18P	8P	RLS	INB	BSV79	2N4860	0
2N4857A	NJD	B6	40V	100MA	200C	1.8WC	2/6V	20/100MA		40R	10P	4P	RLS	TDY	BSV78	2N4857	0

TYPE NO	CONS TRUC TION	CASE & LEAD	V DS MAX	I D MAX	T J MAX	P TOT MAX	VP OR VT	IDSS OR IDOM	G MO	R DS MAX	C ISS MAX	C RSS MAX	USE	SUPP LIER	EUR SUB	USA SUB	ISS
2N4857CHP	NJD	B73	40V	100MA	200C		2/6V	20/100MA		40R	18P	8P	RMS	SIU			0
2N4857JAN	NJD	B6	40V	100MA	200C	.36WF	2/6V	20/100MA		40R	18P	8P	RMS	SIU	BSV79	2N4091	0
2N4857T092	NJD	B63	40V	250MA	125C	1.8WF	2/6V	20/100MA		40R	18P	8P	RLS	INB	BSV79	2N4860	0
2N4858	NJD	B6	40V	250MA	200C	1.8WC	0.8/4V	8/80MA		60R	18P	8P	RLS	INB	BSV80	2N4861	0
2N4858/D	NJD	B73	40V	250MA	200C	1.8WC	0.8/4V	8/80MA		60R	18P	8P	RLS	INB	BSV80	2N4858	0
2N4858/W	NJD	B74	40V	250MA	200C		0.8/4V	8/80MA		60R	18P	8P	RLS	INB	BSV80	2N4861	0
2N4858A	NJD	B6	40V	80MA	200C	1.8WC	0.8/4V	8/80MA		60R	10P	4P	RLS	TDY	BSV79	2N4858	0
2N4858CHP	NJD	B73	40V	80MA	200C		0.8/4V	8/80MA		60R	18P	8P	RMS	SIU			0
2N4858JAN	NJD	B6	40V	100MA	200C	.36WF	0.8/4V	8/80MA		60R	18P	8P	RMS	SIU	BSV80	2N4092	0
2N4858T092	NJD	B63	40V	80MA	125C	1.8WC	0.8/4V	8/80MA		60R	18P	8P	RLS	INB	BSV80	2N4858	0
2N4859	NJD	B6	30V	250MA	200C	1.8WC	4/10V	50MNMA		25R	18P	8P	RLS	INB	BSV78	2N4856	0
2N4859/D	NJD	B73	30V	250MA	200C		4/10V	50MNMA		25R	18P	8P	RLS	INB	BSV78	2N4856	0
2N4859/W	NJD	B74	30V	250MA	200C		4/10V	50MNMA		25R	18P	8P	RLS	INB	BSV78	2N4856	0
2N4859A	NJD	B6	30V	200MA	200C	1.8WC	4/10V	50MNMA		25R	10P	4P	RLS	TDY	BSV78	2N4859	0
2N4859CHP	NJD	B73	40V	250MA	200C		4/10V	50MNMA		25R	18P	8P	RMS	SIU			0
2N4859JAN	NJD	B6	30V	100MA	200C	.36WF	4/10V	50MNMA		25R	18P	8P	RMS	SIU	BSV78	2N4391	0
2N4859T092	NJD	B63	30V	250MA	125C	1.8WC	4/10V	50MNMA		25R	18P	8P	RLS	INB	BSV78	2N4856	0
2N4860	NJD	B6	30V	250MA	200C	1.8WC	2/6V	20/100MA		40R	18P	8P	RLS	INB	BSV79	2N4857	0
2N4860/D	NJD	B73	30V	250MA			2/6V	20/100MA		40R	18P	8P	RLS	INB	BSV79	2N4857	0
2N4860/W	NJD	B74	30V	250MA	200C		2/6V	20/100MA		40R	18P	8P	RLS	INB	BSV79	2N4857	0
2N4860A	NJD	B6	30V	100MA	200C	1.8WC	2/6V	20/100MA		40R	10P	4P	RLS	TDY	BSV79	2N4860	0
2N4860CHP	NJD	B73	40V	100MA	200C		2/6V	20/100MA		40R	18P	8P	RMS	SIU			0
2N4860JAN	NJD	B6	30V	100MA	200C	.36WF	2/6V	20/100MA		40R	18P	8P	RMS	SIU	BSV79	2N4091	0
2N4860T092	NJD	B63	30V	250MA	125C	1.8WC	2/6V	20/100MA		40R	18P	8P	RLS	INB	BSV79	2N4857	0
2N4861	NJD	B6	30V	250MA	200C	1.8WC	0.8/4V	8/80MA		60R	18P	8P	RLS	INB	BSV80	2N4858	0

TYPE NO	CONS TRUC TION	CASE & LEAD	V DS MAX	I D MAX	T J MAX	P TOT MAX	VP OR VT	IDSS OR IDOM	G MO	R DS MAX	C ISS MAX	C RSS MAX	USE	SUPP LIER	EUR SUB	USA SUB	ISS
2N4861/D	NJD	B73	30V	250MA	200C		0.8/4V	8/80MA		60R	18P	8P	RLS	INB	BSV80	2N4858	0
2N4861/W	NJD	B74	30V	250MA	200C		0.8/4V	8/80MA		60R	18P	8P	RLS	INB	BSV80	2N4858	0
2N4861A	NJD	B6	30V	80MA	200C	1.8WC	0.8/4V	8/80MA		60R	10P	4P	RLS	TDY	BSV79	2N4861	0
2N4861CHP	NJD	B73	40V	80MA	200C		0.8/4V	8/80MA		60R	18P	8P	RMS	SIU			0
2N4861JAN	NJD	B6	30V	100MA	200C	.36WF	0.8/4V	8/80MA		60R	18P	8P	RMS	SIU	BSV80	2N4392	0
2N4861T092	NJD	B63	30V	250MA	125C	1.8WC	0.8/4V	8/80MA		60R	18P	8P	RLS	INB	BSV80	2N4858	0
2N4867	NJD	B15	40V	50MA	175C	0.3WF	2MXV	1.2MXMA	0.07/2MO		25P		ALN	SIU	BF800	2N3686A	0
2N4867/D	NJD	B73	40V	8MA	200C	0.3WF	0.7/2V	.4/1.2MA	0.7/2MO		25P	5P	ALN	INB	BF800	2N4867A	0
2N4867/W	NJD	B74	40V	8MA	200C	0.3WF	0.7/2V	.4/1.2MA	0.7/2MO		25P	5P	ALN	INB	BF800	2N4867A	0
2N4867A	NJD	B15	40V	8MA	200C	0.3WF	0.7/2V	.4/1.2MA	0.7/2MO		25P	5P	ALN	INB	BF800	2N3686A	0
2N4867A/D	NJD	B73	40V	8MA	200C	0.3WF	0.7/2V	.4/1.2MA	0.7/2MO		25P	5P	ALN	INB	BF800	2N4867A	0
2N4867A/W	NJD	B74	40V	8MA	200C	0.3WF	0.7/2V	.4/1.2MA	0.7/2MO		25P	5P	ALN	INB	BF800	2N4867A	0
2N4867AT92	NJD	B63	40V	8MA	125C	0.3WF	0.7/2V	.4/1.2MA	0.7/2MO		25P	5P	ALN	INB	BF800	2N4867A	0
2N4867CHP	NJD	B73	40V	2MA	200C		0.7/2V	.4/1.2MA	0.7/2MO		25P	5P	ALN	SIU			0
2N4867T092	NJD	B63	40V	8MA	125C	0.3WF	0.7/2V	.4/1.2MA	0.7/2MO		25P	5P	ALN	INB	BF800	2N4867A	0
2N4868	NJD	B15	40V	8MA	200C	0.3WF	1/3V	1/3MA	1/3MO		25P	5P	ALG	INB	BF808	2N4868A	0
2N4868/D	NJD	B73	40V	8MA	200C	0.3WF	1/3V	1/3MA	1/3MO		25P	5P	ALN	INB	BF808	2N4868A	0
2N4868/W	NJD	B74	40V	8MA	200C	0.3WF	1/3V	1/3MA	1/3MO		25P	5P	ALN	INB	BF808	2N4868A	0
2N4868A	NJD	B15	40V	8MA	200C	0.3WF	1/3V	1/3MA	1/3MO		25P	5P	ALN	INB	BF808	2N3685A	0
2N4868A/D	NJD	B73	40V	8MA	200C	0.3WF	1/3V	1/3MA	1/3MO		25P	5P	ALN	INB	BF808	2N4868A	0
2N4868A/W	NJD	B74	40V	8MA	200C	0.3WF	1/3V	1/3MA	1/3MO		25P	5P	ALN	INB	BF808	2N4868A	0
2N4868AT92	NJD	B63	40V	8MA	125C	0.3WF	1/3V	1/3MA	1/3MO		25P	5P	ALN	INB	BF808	2N4868A	0
2N4868CHP	NJD	B73	40V	3MA	200C		1/3V	1/3MA	1/3MO		25P	5P	ALN	SIU			0
2N4868T092	NJD	B63	40V	8MA	125C	0.3WF	1/3V	1/3MA	1/3MO		25P	5P	ALN	INB	BF808	2N4868A	0
2N4869	NJD	B15	40V	8MA	200C	0.3WF	1.8/5V	2/7.5MA	1.3/4MO		25P	5P	ALN	INB	BFW56	2N4869A	0

TYPE NO	CONS TRUC TION	CASE & LEAD	V DS MAX	I D MAX	T J MAX	P TOT MAX	VP OR VT	IDSS OR IDOM	G MO	R DS MAX	C ISS MAX	C RSS MAX	USE	SUPP LIER	EUR SUB	USA SUB	ISS
2N4869/D	NJD	B73	40V	8MA	200C	0.3WF	1.8/5V	2/7.5MA	1.3/4MO		25P	5P	ALN	INB	BFW56	2N4869A	0
2N4869/W	NJD	B74	40V	8MA	200C	0.3WF	1.8/5V	2/7.5MA	1.3/4MO		25P	5P	ALN	INB	BFW56	2N4869A	0
2N4869A	NJD	B15	40V	8MA	200C	0.3WF	1.8/5V	2/7.5MA	1.3/4MO		25P	5P	ALN	INB	BFW56	2N3684A	0
2N4869A/D	NJD	B73	40V	8MA	200C	0.3WF	1.8/5V	2/7.5MA	1.3/4MO		25P	5P	ALN	INB	BFW56	2N4869A	0
2N4869A/W	NJD	B74	40V	8MA	200C	0.3WF	1.8/5V	2/7.5MA	1.3/4MO		25P	5P	ALN	INB	BFW56	2N4869A	0
2N4869AT92	NJD	B63	40V	8MA	125C	0.3WF	1.8/5V	2/7.5MA	1.3/4MO		25P	5P	ALN	INB	BFW56	2N4869A	0
2N4869CHP	NJD	B73	40V	8MA	200C		1.8/5V	2.5/7MA	1.3/4MO		25P	5P	ALN	SIU			0
2N4869T092	NJD	B63	40V	8MA	125C	0.3WF	1.8/5V	2/7.5MA	1.3/4MO		25P	5P	ALN	INB	BFW56	2N4869A	0
2N4881	NJD	B22	300V	2MA	200C	0.8WF	0.5/15V	0.4/2MA	0.35/1MO	5K	10P	1P5	ALH	TDY		2N5543	0
2N4882	NJD	B22	300V	8MA	200C	0.8WF	0.5/15V	1.5/8MA	0.6/1.5MO	3K	15P	1P5	ALH	TDY		2N5543	0
2N4883	NJD	B22	200V	2MA	200C	0.8WC	0.5/10V	0.4/2MA	0.35/1MO		15P	1P5	ALH	TDY		2N4881	0
2N4884	NJD	B22	200V	8MA	200C	0.8WF	0.5/10V	1.5/8MA	0.6/1.5MO		15P	1P5	ALH	TDY		2N4882	0
2N4885	NJD	B22	125V	2MA	200C	0.8WF	0.5/10V	0.4/2MA	0.35/1MO		15P	1P5	ALH	TDY		2N4883	0
2N4886	NJD	B22	125V	8MA	200C	0.8WF	0.5/10V	1.5/8MA	0.6/1.5MO		15P	1P5	ALH	TDY		2N4884	0
2N4977	NJD	B6	30V	250MA	200C	1.8WC	4/10V	50MNMA		15R	35P	8P	ALC	TDY		2N5434	0
2N4978	NJD	B6	30V	100MA	200C	1.8WC	2/8V	15MNMA		20R	35P	8P	ALC	TDY		2N5434	0
2N4979	NJD	B6	30V	8MA	200C	1.8WC	0.5/5V	7.5MNMA		40R	35P	8P	ALC	TDY		2N4857	0
2N5018	PJD	B7	30V	50MA	200C	1.8WC	10MXV	10MNMA		75R	45P	10P	RLS	TDY		2N5114	0
2N5018CHP	PJD	B73	30V	90MA	200C		10MXV	10MNMA		75R	25P	7P	RLS	SIU			0
2N5019	PJD	B7	30V	50MA	200C	1.8WC	5MXV	5MNMA		150R	45P	10P	RLS	TDY		2N5116	0
2N5019CHP	PJD	B73	30V	90MA	200C		5MXV	5MNMA	1	50R	25P	7P	RLS	SIU			0
2N5020	PJD	B7	25V	2MA	175C	0.3WF	0.3/1.5V	.3/1.2MA	1/3.5MO	1K3	25P	7P	RLG	TDY	BF320A	2N5266	0
2N5021	PJD	B7	25V	4MA	175C	0.3WF	0.5/2.5V	1/3.5MA	1.5/5MO	1K	25P	7P	RLG	TDY	BF320A	2N2608	0
2N5033	PJD	B10	20V	3.5MA	150C	.22WF	0.3/2.5V	.3/3.5MA	1/5MO	1K3	25P	7P	RLG	TDY	BF320A	2N5460	0
2N5045	NJD	B20	50V	8MA	175C	0.4WF	0.5/4.5V	0.5/8MA	1.5/6MO		8P	4P	DUA	SIU	BFQ10	2N5196	0

TYPE NO	CONS TRUC TION	CASE & LEAD	V DS MAX	I D MAX	T J MAX	P TOT MAX	VP OR VT	IDSS OR IDOM	G MO	R DS MAX	C ISS MAX	C RSS MAX	USE	SUPP LIER	EUR SUB	USA SUB	ISS
2N5045X2	NJD	B20	50V	8MA	175C	0.4WF	0.5/4.5V	0.5/8MA	1.5/6MO		8P	4P	DUA	SIU	BFQ10	2N5196	0
2N5046	NJD	B20	50V	8MA	175C	0.4WF	0.5/4.5V	0.5/8MA	1.5/6MO		8P	4P	DUA	SIU	BFQ10	2N5196	0
2N5046X2	NJD	B20	50V	8MA	175C	0.4WF	0.5/4.5V	0.5/8MA	1.5/6MO		8P	4P	DUA	SIU	BFQ10	2N5196	0
2N5047	NJD	B20	50V	8MA	175C	0.4WF	0.5/4.5V	0.5/8MA	1.5/6MO		8P	4P	DUA	SIU	BFQ10	2N5196	0
2N5047CHP	NJD	B73	50V	8MA	200C		0.5/4.5V	0.5/8MA	1.5/6.5MO		8P	4P	DUA	SIU			0
2N5047X2	NJD	B20	50V	8MA	175C	0.4WF	0.5/4.5V	0.5/8MA	1.5/6MO		8P	4P	DUA	SIU	BFQ10	2N5196	0
2N5078	NJD	B15	30V	25MA	200C	0.3WF	0.5/8V	4/25MA	4.5/10MO		6P	2P	FVG	TDY	BF256B	2N4416	0
2N5103	NJD	B15	25V	8MA	200C	0.3WF	0.5/4V	1/8MA	2/8MO		5P	1P	FVG	TDY	BF256A	2N5246	0
2N5104	NJD	B15	25V	15MA	200C	0.3WF	0.5/4V	2/6MA	3.5/7.5MO		5P	1P	FVG	TDY	BF256A	2N5246	0
2N5105	NJD	B15	25V	15MA	200C	0.3WF	0.5/4V	5/15MA	5/10MO		5P	1P	FVG	TDY	BF256B	2N4416	0
2N5114	PJD	B7	30V	90MA	200C	0.5WF	5/10V	30/90MA		75R	25P	7P	RLS	INB		2N5114	0
2N5114/D	PJD	B73	30V	90MA	200C	0.5WF	5/10V	30/90MA		75R	25P	7P	RLS	INB		2N5114	0
2N5114/W	PJD	B74	30V	90MA	200C	0.5WF	5/10V	30/90MA		75R	25P	7P	RLS	INB		2N5114	0
2N5114CHP	PJD	B73	30V	90MA	200C		5/10V	30/90MA		75R	25P	7P	RLS	SIU			0
2N5114JAW	PJD	B7	30V	90MA	200C	0.5WF	5/10V	30/90MA		75R	25P	7P	RLS	INB		2N5114	0
2N5114T092	PJD	B63	30V	90MA	125C	0.5WF	5/10V	30/90MA		75R	25P	7P	RLS	INB		2N5114	0
2N5115	PJD	B7	30V	60MA	200C	0.5WF	3/6V	15/60MA		100R	25P	7P	RLS	CRY		2N5114	0
2N5115/D	PJD	B73	30V	90MA	200C	0.5WF	3/6V	15/60MA		100R	25P	7P	RLS	INB		2N5115	0
2N5115/W	PJD	B74	30V	90MA	200C	0.5WF	3/6V	15/60MA		100R	25P	7P	RLS	INB		2N5115	0
2N5115CHP	PJD	B73	30V	90MA	200C		3/6V	15/60MA		100R	25P	7P	RLS	SIU			0
2N5115JAN	PJD	B7	30V	90MA	200C	0.5WF	3/6V	15/60MA		100R	25P	7P	RLS	INB		2N5115	0
2N5115T092	PJD	B63	30V	90MA	125C	0.5WF	3/6V	15/60MA		100R	25P	7P	RLS	INB		2N5115	0
2N5116	PJD	B7	30V	25MA	200C	0.5WF	1/4V	5/25MA		150R	25P	7P	RLS	CRY		2N3382	0
2N5116/D	PJD	B73	30V	90MA	200C	0.5WF	1/4V	5/25MA		150R	25P	7P	RLS	INB		2N3993	0
2N5116/W	PJD	B74	30V	90MA	200C	0.5WF	1/4V	5/25MA		150R	25P	7P	RLS	INB		2N3993	0

TYPE NO	CONS TRUC TION	CASE & LEAD	V DS MAX	I D MAX	T J MAX	P TOT MAX	VP OR VT	IDSS OR IDOM	G MO	R DS MAX	C ISS MAX	C RSS MAX	USE	SUPP LIER	EUR SUB	USA SUB	ISS
2N5116CHP	PJD	B73	30V	90MA	200C		1/4V	5/25MA		150R	25P	7P	RLS	SIU			0
2N5116JAN	PJD	B7	30V	90MA	200C	0.5WF	1/4V	5/25MA		175R	27P	7P	RLS	INB		2N3993	0
2N5116T092	PJD	B63	30V	90MA	125C	0.5WF	1/4V	5/25MA		150R	25P	7P	RLS	INB		2N3993	0
2N5158	NJD	B47	40V	400MA	200C	0.4WF	8MXV	30/90MA			50P		ALC	CRY	BSV78	2N5434	0
2N5159	NJD	B47	40V	400MA	200C	0.4WF	6MXV	30/90MA			50P		ALC	CRY	BSV78	2N5433	0
2N5163	NJD	B10	25V	40MA	125C	0.3WF	0.4/8V	1/40MA	2/9MO		20P	5P	FVG	TDY	BFW61	2N5398	0
2N5196	NJD	B51	50V	7MA	150C	0.5WF	0.7/4V	0.7/7MA	1/4MO		6P	2P	DUA	INB		2N5545	0
2N5196/D	NJD	B73	50V	7MA	150C	0.5WF	0.7/4V	0.7/7MA	1/4MO		6P	2P	DUA	INB		2N5545	0
2N5196/W	NJD	B74	50V	7MA	150C	0.5WF	0.7/4V	0.7/7MA	1/4MO		6P	2P	DUA	INB		2N5545	0
2N5196X2	NJD	B51	50V	7MA	150C	0.5WF	0.7/4V	0.7/7MA	1/4MO		6P	2P	DUA	INB		2N5545	0
2N5197	NJD	B51	50V	7MA	150C	0.5WF	0.7/4V	0.7/7MA	1/4MO		6P	2P	DUA	INB		2N5545	0
2N5197/D	NJD	B73	50V	7MA		0.5WF	0.7/4V	0.7/7MA	1/4MO		6P	2P	DUA	INB		2N5545	0
2N5197/W	NJD	B74	50V	7MA	150C	0.5WF	0.7/4V	0.7/7MA	1/4MO		6P	2P	DUA	INB		2N5196	0
2N5197X2	NJD	B51	50V	7MA	150C	0.5WF	0.7/4V	0.7/7MA	1/4MO		6P	2P	DUA	INB		2N5196	0
2N5198	NJD	B51	50V	7MA	150C	0.5WF	0.7/4V	0.7/7MA	1/4MO		6P	2P	DUA	INB		2N5196	0
2N5198/D	NJD	B73	50V	7MA	150C	0.5WF	0.7/4V	0.7/7MA	1/4MO		6P	2P	DUA	INB		2N5196	0
2N5198/W	NJD	B74	50V	7MA	150C	0.5WF	0.7/4V	0.7/7MA	1/4MO		6P	2P	DUA	INB		2N5196	0
2N5198X2	NJD	B51	50V	7MA	150C	0.5WF	0.7/4V	0.7/7MA	1/4MO		6P	2P	DUA	INB		2N5196	0
2N5199	NJD	B51	50V	7MA	150C	0.5WF	0.7/4V	0.7/7MA	1/4MO		6P	2P	DUA	INB		2N5196	0
2N5199/D	NJD	B73	50V	7MA	200C	0.5WF	0.7/4V	0.7/7MA	1/4MO		6P	2P	DUA	INB		2N5196	0
2N5199/W	NJD	B74	50V	7MA	200C	0.5WF	0.7/4V	0.7/7MA	1/4MO		6P	2P	DUA	INB		2N5196	0
2N5199CHP	NJD	B73	50V	7MA	200C		0.7/4V	0.7/7MA	1/4MO		6P	2P	DUA	SIU			0
2N5199X2	NJD	B51	50V	7MA	150C	0.5WF	0.7/4V	0.717MA	1/4MO		6P	2P	DUA	INB		2N5196	0
2N5245	NJD	B10	30V	15MA	150C	.36WF	1/6V	5/15MA	4MNMO		4P5	1P2	FVG	TDY	BF256B	2N5245	0
2N5246	NJD	B10	30V	7MA	150C	.36WF	0.5/4V	1.5/7MA	2.5MNMO		4P5	1P2	FVG	TDY	BF256A	2N5485	0

TYPE NO	CONS TRUC TION	CASE & LEAD	V DS MAX	I D MAX	T J MAX	P TOT MAX	VP OR VT	IDSS OR IDOM	G MO	R DS MAX	C ISS MAX	C RSS MAX	USE	SUPP LIER	EUR SUB	USA SUB	ISS
2N5247	NJD	B10	30V	25MA	150C	.36WF	1.5/8V	8/24MA	4MNMO		4P5	1P2	FVG	TDY	BF256C	2N5247	0
2N5248	NJD	B1	30V	20MA	125C	0.3WF	8MXV	20MXMA	3.5/6.5MO		6P		FVG	TIB	BF244	2N3823	0
2N5265	PJD	B15	60V	20MA	175C	0.3WF	3MXV	0.5/1MA	0.9/2.7MO		7P	2P	ALG	INB	BF320A	2N2607	0
2N5266	PJD	B15	60V	20MA	175C	0.3WF	3MXV	.8/1.6MA	1/3MO		7P	2P	ALG	INB	BF320A	2N2844	0
2N5267	PJD	B15	60V	20MA	175C	0.3WF	6MXV	1.5/3MA	1.5/3.5MO		7P	2P	ALG	INB	BF320A	2N3330	0
2N5268	PJD	B15	60V	20MA	175C	0.3WF	6MXV	2.5/5MA	2/4MO		7P	2P	ALG	INB	BF320B	2N3378	0
2N5269	PJD	B15	60V	20MA	175C	0.3WF	8MXV	4/8MA	2.2/4.5MO		7P	2P	ALG	INB	BF320B	2N3378	0
2N5270	PJD	B15	60V	20MA	175C	0.3WF	8MXV	7/14MA	2.5/5MO		7P	2P	ALG	INB	BF320C	2N3382	0
2N5277	NJD	B22	150V	13MA	200C	0.8WF	0.5/7V	2.5/13MA	2/5MO		25P	2P	ALH	TDY		2N5543	0
2N5277CHP	NJD	B73	150V	25MA	200C		0.7/7V	2.5/12MA	2/5MO		25P	4P	ALH	SIU			0
2N5278	NJD	B22	150V	25MA	200C	0.8WF	2/10V	10/25MA	3/6MO		25P	5P	ALH	TDY		2N5543	0
2N5278CHP	NJD	B73	150V	25MA	200C		2/10V	10/25MA	3/6MO		25P	4P	ALH	SIU			0
2N5358	NJD	B14	40V	3MA	175C	0.3WF	0.5/3V	0.5/1MA	1/3MO		6P	2P	ALN	SIU	BFW13	2N4867A	0
2N5358CHP	NJD	B73	40V	8MA	200C		1/3V	0.5/1MA	1/3MO		6P	2P	RLG	SIU			0
2N5359	NJD	B14	40V	3MA	175C	0.3WF	0.8/4V	.8/1.6MA	1.2/3.6MO		6P	2P	ALN	SIU	BF347	2N3685A	0
2N5359CHP	NJD	B73	40V	18MA	200C		0.8/4V	.8/1.6MA	1.2/3.6MO		6P	2P	RLG	SIU			0
2N5360	NJD	B14	40V	3MA	175C	0.3WF	0.8/4V	1.5/3MA	1.4/4.2MO		6P	2P	ALN	SIU	BC264A	2N4868A	0
2N5360CHP	NJD	B73	40V	18MA	200C		0.8/4V	1.5/3MA	1.4/4.3MO		6P	2P	RLG	SIU			0
2N5361	NJD	B14	40V	18MA	175C	0.3WF	1/6V	2.5/5MA	1.5/4.5MO		6P	2P	ALN	SIU	BC264A	2N4869A	0
2N5361CHP	NJD	B73	40V	18MA	200C		1.6/6V	2.5/5MA	1.5/4.5MO		6P	2P	RLG	SIU			0
2N5362	NJD	B14	40V	18MA	175C	0.3WF	2/7V	4/8MA	2/5MO		6P	2P	ALN	SIU	BFW11	2N3684A	0
2N5362CHP	NJD	B73	40V	18MA	200C		2/7V	4/8MA	2/5.5MO		6P	2P	RLG	SIU			0
2N5363	NJD	B14	40V	18MA	175C	0.3WF	2.5/8V	7/14MA	2.5/6MO		6P	2P	ALN	SIU	BC264D	2N4303	0
2N5363CHP	NJD	B73	40V	18MA	200C		2.5/8V	7/14MA	2.5/6MO		6P	2P	RLG	SIU			0
2N5364	NJD	B14	40V	18MA	175C	0.3WF	2.5/8V	9/18MA	2.7/6.5MO		6P	2P	ALN	SIU	BFW10	2N5245	0

TYPE NO	CONS TRUC TION	CASE & LEAD	V DS MAX	I D MAX	T J MAX	P TOT MAX	VP OR VT	IDSS OR IDOM	G MO	R DS MAX	C ISS MAX	C RSS MAX	USE	SUPP LIER	EUR SUB	USA SUB	ISS
2N5364CHP	NJD	B73	40V	18MA	200C		2.5/8V	9/18MA	2.5/6.5MO		6P	2P	RLG	SIU			0
2N5391	NJD	B6	70V	2MA	200C	0.3WF	0.5/2V	1.5MXMA	1.5/4.5MO		18P	5P	ALN	TDY	BF800	2N5392	0
2N5392	NJD	B6	70V	3MA	200C	0.3WF	0.5/2.5V	1/3MA	2/6MO		18P	5P	ALN	TDY	BF808	2N5393	0
2N5393	NJD	B6	70V	5MA	200C	0.3WF	1/3V	2.5/5MA	3/6.5MO		18P	5P	ALN	TDY	BFW56	2N5394	0
2N5394	NJD	B6	70V	6MA	200C	0.3WF	1/4V	4/6MA	4/7.5MO		18P	5P	ALN	TDY	BF810	2N5395	0
2N5395	NJD	B6	70V	8MA	200C	0.3WF	1.5/4V	5.5/8MA	4.5/7.5MO		18P	5P	ALN	TDY	BF805	2N5396	0
2N5396	NJD	B6	70V		200C	0.3WF	2/5V	7.5/10MA	4.5/7.5MO		18P	5P	ALN	TDY	BF805	2N3822	0
2N5397	NJD	B15	25V	30MA	200C	0.3WF	1/6V	10/30MA	6/10MO		5P	1P2	TUG	INB	BF256C	2N5247	0
2N5397/D	NJD	B73	25V	30MA	200C	0.3WF	1/6V	10/30MA	6/10MO		5P	1P2	TUG	INB	BF256C	2N5247	0
2N5397/W	NJD	B74	25V	30MA	200C	0.3WF	1/6V	10/30MA	6/10MO		5P	1P2	TUG	INB	BF256C	2N5397	0
2N5397CHP	NJD	B73	25V	40MA	200C		1/6V	10/30MA	6/10MO		5P	1P2	ULG	SIU			0
2N5397T092	NJD	B63	25V	30MA	125C	0.3WF	1/6V	10/30MA	6/10MO		5P	1P2	TUG	INB	BF256C	2N5247	0
2N5398	NJD	B15	25V	40MA	200C	0.3WF	1/6V	5/40MA	5.5/10MO		5P5	1P3	TUG	INB	BF256C	2N5245	0
2N5398/D	NJD	B73	25V	40MA		0.3WF	1/6V	5/40MA	5.5/10MO		5P5	1P3	TUG	INB	BF256B	2N5247	0
2N5398/W	NJD	B74	25V	40MA	200C	0.3WF	1/6V	5/40MA	5.5/10MO		5P5	1P3	TUG	INB	BF256B	2N5245	0
2N5398T092	NJD	B63	25V	40MA	125C	0.3WF	1/6V	5/40MA	5.5/10MO		5P5	1P3	TUG	INB	BF256B	2N5247	0
2N5432	NJD	B6	25V	400MA	200C	0.3WF	4/10V	150MNMA		5R	30P	15P	RMS	INB		2N5432	0
2N5432/D	NJD	B73	25V	400MA	200C	0.3WF	4/10V	150MNMA		5R	30P	15P	RMS	INB		2N5432	0
2N5432/W	NJD	B74	25V	400MA	200C	0.3WF	4/10V	150MNMA		5R	30P	15P	RMS	INB		2N5432	0
2N5432CHP	NJD	B73	25V	500MA	200C		4/10V	150MNMA		5R	30P	15P	RMS	SIU			0
2N5432T092	NJD	B63	25V	400MA	125C	0.3WF	4/10V	150MNMA		5R	30P	15P	RMS	INB		2N5432	0
2N5433	NJD	B6	25V	400MA	200C	0.3WF	3/9V	100MNMA		7R	30P	15P	RMS	INB		2N5432	0
2N5433/D	NJD	B73	25V	400MA	200C	0.3WF	3/9V	100MNMA		7R	30P	15P	RMS	INB		2N5432	0
2N5433/W	NJD	B74	25V	400MA	200C	0.3WF	3/9V	100MNMA		7R	30P	15P	RMS	INB		2N5432	0
2N5433CHP	NJD	B73	25V	500MA	200C		3/9V	100MNMA		7R	30P	15P	RMS	SIU			0

TYPE NO	CONS TRUC TION	CASE & LEAD	V DS MAX	I D MAX	T J MAX	P TOT MAX	VP OR VT	IDSS OR IDOM	G MO	R DS MAX	C ISS MAX	C RSS MAX	USE	SUPP LIER	EUR SUB	USA SUB	ISS
2N5433T092	NJD	B63	25V	400MA	125C	0.3WF	3/9V	100MNMA		7R	30P	15P	RMS	INB		2N5432	0
2N5434	NJD	B6	25V	400MA	200C	0.3WF	1/4V	30MNMA		10R	30P	15P	RMS	INB		2N5433	0
2N5434/D	NJD	B73	25V	400MA	200C	0.3WF	1/4V	30MNMA		10R	30P	15P	RMS	INB		2N5433	0
2N5434/W	NJD	B74	25V	400MA	200C	0.3WF	1/4V	30MNMA		10R	30P	15P	RMS	INB		2N5433	0
2N5434CHP	NJD	B73	25V	500MA	200C		1/4V	30MNMA		10R	30P	15P	RMS	SIU			0
2N5434T092	NJD	B63	25V	400MA	125C	0.3WF	1/4V	30MNMA		10R	30P	15P	RMS	INB		2N5433	0
2N5452	NJD	B51	50V	5MA	150C	0.5WF	1/4.5V	0.5/5MA	1/3MO		4P	1P2	DUA	INB	BFQ10	2N3954	0
2N5452X2	NJD	B51	50V	5MA	150C	0.5WF	1/4.5V	0.5/5MA	1/3MO		4P	1P2	DUA	INB	BFQ10	2N3954	0
2N5453	NJD	B51	50V	5MA	150C	0.5WF	1/4.5V	0.5/5MA	1/3MO		4P	1P2	DUA	INB	BFQ10	2N5452	0
2N5453X2	NJD	B51	50V	5MA	150C	0.5WF	1/4.5V	0.5/5MA	1/3MO		4P	1P2	DUA	INB	BFQ10	2N5452	0
2N5454	NJD	B51	50V	5MA	150C	0.5WF	1/4.5V	0.5/5MA	1/3MO		4P	1P2	DUA	INB	BFQ10	2N5453	0
2N5454X2	NJD	B51	50V	5MA	150C	0.5WF	1/4.5V	0.5/5MA	1/3MO		4P	1P2	DUA	INB	BFQ10	2N5453	0
2N5457	NJD	B3	25V	16MA	135C	.31WF	0.5/6V	1/5MA	1/5MO		7P	3P	ALG	INB	BFW12	2N4868A	0
2N5458	NJD	B3	25V	16MA	135C	.31WF	1/7V	2/9MA	1.5/5.5MO		7P	3P	ALG	INB	BFW61	2N3822	0
2N5459	NJD	B3	25V	16MA	135C	.31WF	2/8V	4/16MA	2/6MO		7P	3P	ALG	INB	BC264C	2N4303	0
2N5460	PJD	B63	40V	16MA	135C	.31WF	0.75/6V	1/5MA	1/4MO		7P	2P	ALN	INB	BF320A	2N3332	0
2N5460/D	PJD	B73	40V	16MA	135C	.31WF	0.75/6V	1/5MA	1/4MO		7P	2P	ALN	INB	BF320A	2N3332	0
2N5460/W	PJD	B74	40V	16MA	135C	.31WF	0.75/6V	1/5MA	1/4MO		7P	2P	ALN	INB	BF320A	2N3332	0
2N5461	PJD	B63	40V	16MA	135C	.31WF	1/7.5V	2/9MA	1.5/5MO		7P	2P	ALN	INB	BF320B	2N3378	0
2N5461/D	PJD	B73	40V	16MA	135C	.31WF	1/7.5V	2/9MA	1.5/5MO		7P	2P	ALN	INB	BF320B	2N3378	0
2N5461/W	PJD	B74	40V	16MA	135C	.31WF	1/7.5V	2/9MA	1.5/5MO		7P	2P	ALN	INB	BF320B	2N3378	0
2N5462	PJD	B63	40V	16MA	135C	.31WF	1.8/9V	4/16MA	2/6MO		7P	2P	ALN	INB	BF320B	2N3330	0
2N5462/D	PJD	B73	40V	16MA	135C	.31WF	1.8/9V	4/16MA	2/6MO		7P	2P	ALN	INB	BF320B	2N3330	0
2N5462/W	PJD	B74	40V	16MA	135C	.31WF	1.8/9V	4/16MA	2/6MO		7P	2P	ALN	INB	BF320B	2N3330	0
2N5463	PJD	B63	60V	16MA	135C	.31WF	0.75/6V	1/5MA	1/4MO		7P	2P	ALN	INB	BF320A	2N3332	0

TYPE NO	CONS TRUC TION	CASE & LEAD	V DS MAX	I D MAX	T J MAX	P TOT MAX	VP OR VT	IDSS OR IDOM	G MO	R DS MAX	C ISS MAX	C RSS MAX	USE	SUPP LIER	EUR SUB	USA SUB	ISS
2N5463/D	PJD	B73	60V	16MA	135C	.31WF	0.75/6V	1/5MA	1/4MO		7P	2P	ALN	INB	BF320A	2N3332	0
2N5463/W	PJD	B74	60V	16MA	135C	.31WF	0.75/6V	1/5MA	1/4MO		7P	2P	ALN	INB	BF320A	2N3332	0
2N5464	PJD	B63	60V	16MA	135C	.31WF	1/7.5V	2/9MA	1.5/5MO		7P	2P	ALN	INB	BF320B	2N3330	0
2N5464/D	PJD	B73	60V	16MA	135C	.31WF	1/7.5V	2/9MA	1.5/5MO		7P	2P	ALN	INB	BF320B	2N3330	0
2N5464/W	PJD	B74	60V	16MA	135C	.31WF	1/7.5V	2/9MA	1.5/5MO		7P	2P	ALN	INB	BF320B	2N3330	0
2N5465	PJD	B63	60V	16MA	135C	.31WF	1.8/9V	4/16MA	2/6MO		7P	2P	ALN	INB	BF320B	2N5269	0
2N5465/D	PJD	B73	60V	16MA	135C	.31WF	1.8/9V	4/16MA	2/6MO		7P	2P	ALN	INB	BF320B	2N5269	0
2N5465/W	PJD	B74	60V	16MA	135C	.31WF	1.8/9V	4/16MA	2/6MO		7P	2P	ALN	INB	BF320B	2N5269	0
2N5471	PJD	B12	40V	2MA	175C	0.3WF	0.5/4V	0.06MXMA	.06/.18MO		5P	1P	ALG	MOB		2N2606	0
2N5472	PJD	B12	40V	2MA	175C	0.3WF	0.7/4V	0.12MXMA	.08/.22MO		5P	1P	ALG	MOB		2N2606	0
2N5473	PJD	B12	40V	2MA	175C	0.3WF	0.9/6V	.1/.25MA	0.12/.3MO		5P	1P	ALG	MOB		2N2606	0
2N5474	PJD	B12	40V	2MA	175C	0.3WF	1.2/7V	0.2/.5MA	0.16/.4MO		5P	1P	ALG	MOB		2N2606	0
2N5475	PJD	B12	40V	2MA	175C	0.3WF	1.5/8V	0.4/1MA	0.2/.5MO		5P	1P	ALG	MOB		2N2843	0
2N5476	PJD	B12	40V	2MA	175C	0.3WF	2/9V	0.8/2MA	.26/.65MO		5P	1P	ALG	MOB		2N2844	0
2N5484	NJD	B3	25V	5MA	150C	.31WF	0.3/3V	1/5MA	2.5MNMO		5P	1P2	FVG	TDY	BF256A	2N5246	0
2N5485	NJD	B3	25V	30MA	150C	.31WF	0.5/4V	4/10MA	3.5/7MO		5P	1P	TUG	INB	BF256B	2N5245	0
2N5486	NJD	B3	25V	30MA	150C	.31WF	2/6V	8/20MA	4/8MO		5P	1P	TUG	INB	BF256C	2N5247	0
2N5505	PJD	B51	30V	7MA	150C	.25WF	4MXV	7MXMA	1/3.5MO		16P		DUA	TDY		2N5510	0
2N5505X2	PJD	B51	30V	7MA	150C	.25WF	4MXV	7MXMA	1/3.5MO		16P		DUA	TDY		2N5510	0
2N5506	PJD	B51	30V	7MA	200C	.25WF	4MXV	7MXMA	1/3.5MO		16P		DUA	TDY		2N5505	0
2N5506X2	PJD	B51	30V	7MA	200C	.25WF	4MXV	7MXMA	1/3.5MO		16P		DUA	TDY		2N5505	0
2N5507	PJD	B51	30V	7MA	200C	.25WF	4MXV	7MXMA	1/3.5MO		16P		DUA	TDY		2N5505	0
2N5507X2	PJD	B51	30V	7MA	200C	.25WF	4MXV	7MXMA	1/3.5MO		16P		DUA	TDY		2N5505	0
2N5508	PJD	B51	30V	7MA	200C	.25WF	4MXV	7MXMA	1/3.5MO		16P		DUA	TDY		2N5505	0
2N5508X2	PJD	B51	30V	7MA	200C	.25WF	4MXV	7MXMA	1/3.5MA		16P		DUA	TDY		2N5505	0

TYPE NO	CONS TRUC TION	CASE & LEAD	V DS MAX	I D MAX	T J MAX	P TOT MAX	VP OR VT	IDSS OR IDOM	G MO	R DS MAX	C ISS MAX	C RSS MAX	USE	SUPP LIER	EUR SUB	USA SUB	ISS
2N5509	PJD	B51	30V	7MA	200C	.25WF	5MXV	7MXMA	1/3.5MO		16P		DUA	TDY		2N5505	0
2N5509X2	PJD	B51	30V	7MA	200C	.25WF	5MXV	7MXMA	1/3.5MO		16P		DUA	TDY		2N5505	0
2N5510	PJD	B51	30V	5MA	200C	.25WF	4MXV	5MXMA	0.5/3MO		16P		DUA	TDY		2N5505	0
2N5510X2	PJD	B51	30V	5MA	200C	.25WF	4MXV	5MXMA	0.5/3MO		16P		DUA	TDY		2N5505	0
2N5511	PJD	B51	30V	5MA	200C	.25WF	4MXV	5MXMA	0.5/3MO		16P		DUA	TDY		2N5510	0
2N5511X2	PJD	B51	30V	5MA	200C	.25WF	4MXV	5MXMA	0.5/3MO		16P		DUA	TDY		2N5510	0
2N5512	PJD	B51	30V	5MA	200C	.25WF	4MXV	5MXMA	0.5/3MO		16P		DUA	TDY		2N5510	0
2N5512X2	PJD	B51	30V	5MA	200C	.25WF	4MXV	5MXMA	0.5/3MO		16P		DUA	TDY		2N5510	0
2N5513	PJD	B51	30V	5MA	200C	.25WF	4MXV	5MXMA	0.5/3MO		16P		DUA	TDY		2N5510	0
2N5513X2	PJD	B51	30V	5MA	200C	.25WF	4MXV	5MXMA	0.5/3MO		16P		DUA	TDY		2N5510	0
2N5514	PJD	B51	30V	5MA	200C	.25WF	4MXV	5MXMA	0.5/3MO		16P		DUA	TDY		2N5510	0
2N5514X2	PJD	B51	30V	5MA	200C	.25WF	4MXV	5MXMA	0.5/3MO		16P		DUA	TDY		2N5510	0
2N5515	NJD	B51	40V	7.5MA	200C	0.5WF	0.7/4V	.5/7.5MA	0.5/1MO		25P	5P	DUA	INB		2N5520	0
2N5515/D	NJD	B73	40V	7.5MA	200C	0.5WF	0.7/4V	.5/7.5MA	0.5/1MO		25P	5P	DUA	INB		2N5520	0
2N5515/W	NJD	B74	40V	7.5MA	200C	0.5WF	0.7/4V	.5/7.5MA	0.5/1MO		25P	5P	DUA	INB		2N5520	0
2N5515X2	NJD	B51	40V	7.5MA	200C	0.5WF	0.7/4V	.5/7.5MA	0.5/1MO		25P	5P	DUA	INB		2N5520	0
2N5516	NJD	B51	40V	7.5MA	200C	0.5WF	0.7/4V	.5/7.5MA	0.5/1MO		25P	5P	DUA	INB		2N5521	0
2N5516/D	NJD	B73	40V	7.5MA	200C	0.5WF	0.7/4V	.5/7.5MA	0.5/1MO		25P	5P	DUA	INB		2N5521	0
2N5516/W	NJD	B74	40V	7.5MA	200C	0.5WF	0.7/4V	.5/7.5MA	0.5/1MO		25P	5P	DUA	INB		2N5521	0
2N5516X2	NJD	B51	40V	7.5MA	200C	0.5WF	0.7/4V	.5/7.5MA	0.5/1MO		25P	5P	DUA	INB		2N5521	0
2N5517	NJD	B51	40V	7.5MA	200C	0.5WF	0.7/4V	.5/7.5MA	0.5/1MO		25P	5P	DUA	INB		2N5522	0
2N5517/D	NJD	B73	40V	7.5MA	200C	0.5WF	0.7/4V	.5/7.5MA	0.5/1MO		25P	5P	DUA	INB		2N5522	0
2N5517/W	NJD	B74	40V	7.5MA	200C	0.5WF	0.7/4V	.5/7.5MA	0.5/1MO		25P	5P	DUA	INB		2N5522	0
2N5517X2	NJD	B51	40V	7.5MA	200C	0.5WF	0.7/4V	.5/7.5MA	0.5/1MO		25P	5P	DUA	INB		2N5522	0
2N5518	NJD	B51	40V	7.5MA	200C	0.5WF	0.7/4V	.5/7.5MA	0.5/1MO		25P	5P	DUA	INB		2N5523	0

TYPE NO	CONS TRUC TION	CASE & LEAD	V DS MAX	I D MAX	T J MAX	P TOT MAX	VP OR VT	IDSS OR IDOM	G MO	R DS MAX	C ISS MAX	C RSS MAX	USE	SUPP LIER	EUR SUB	USA SUB	ISS
2N5518/D	NJD	B73	40V	7.5MA	200C	0.5WF	0.7/4V	.5/7.5MA	0.5/1MO		25P	5P	DUA	INB		2N5523	0
2N5518/W	NJD	B74	40V	7.5MA	200C	0.5WF	0.7/4V	.5/7.5MA	0.5/1MO		25P	5P	DUA	INB		2N5523	0
2N5518X2	NJD	B51	40V	7.5MA	200C	0.5WF	0.7/4V	.5/7.5MA	0.5/1MO		25P	5P	DUA	INB		2N5523	0
2N5519	NJD	B51	40V	7.5MA	200C	0.5WF	0.7/4V	.5/7.5MA	0.5/1MO		25P	5P	DUA	INB		2N5524	0
2N5519/D	NJD	B73	40V	7.5MA	200C	0.5WF	0.7/4V	.5/7.5MA	0.5/1MO		25P	5P	DUA	INB		2N5524	0
2N5519/W	NJD	B74	40V	7.5MA	200C	0.5WF	0.7/4V	.5/7.5MA	0.5/1MO		25P	5P	DUA	INB		2N5524	0
2N5519CHP	NJD	B73	40V	8MA	200C		0.7/4V	.5/7.5MA	1/4MO		25P	5P	DUA	SIU			0
2N5519X2	NJD	B51	40V	7.5MA	200C	0.5WF	0.7/4V	.5/7.5MA	0.5/1MO		25P	5P	DUA	INB		2N5524	0
2N5520	NJD	B51	40V	7.5MA	200C	0.5WF	0.7/4V	.5/7.5MA	0.5/1MO		25P	5P	DUA	INB			0
2N5520/D	NJD	B73	40V	7.5MA	200C	0.5WF	0.7/4V	.5/7.5MA	0.5/1MO		25P	5P	DUA	INB			0
2N5520/W	NJD	B74	40V	7.5MA	200C	0.5WF	0.7/4V	.5/7.5MA	0.5/1MO		25P	5P	DUA	INB			0
2N5520X2	NJD	B51	40V	7.5MA	200C	0.5WF	0.7/4V	.5/7.5MA	0.5/1MO		25P	5P	DUA	INB			0
2N5521	NJD	B51	40V	7.5MA	200C	0.5WF	0.7/4V	.5/7.5MA	0.5/1MO		25P	5P	DUA	INB		2N5520	0
2N5521/D	NJD	B73	40V	7.5MA	200C	0.5WF	0.7/4V	.5/7.5MA	0.5/1MO		25P	5P	DUA	INB		2N5520	0
2N5521/W	NJD	B74	40V	7.5MA	200C	0.5WF	0.7/4V	.5/7.5MA	0.5/1MO		25P	5P	DUA	INB		2N5520	0
2N5521X2	NJD	B51	40V	7.5MA	200C	0.5WF	0.7/4V	.5/7.5MA	0.5/1MO		25P	5P	DUA	INB		2N5520	0
2N5522	NJD	B51	40V	7.5MA	200C	0.5WF	0.7/4V	.5/7.5MA	0.5/1MO		25P	5P	DUA	INB		2N5521	0
2N5522/D	NJD	B73	40V	7.5MA	200C	0.5WF	0.7/4V	.5/7.5MA	0.5/1MO		25P	5P	DUA	INB		2N5521	0
2N5522/W	NJD	B74	40V	7.5MA	200C	0.5WF	0.7/4V	.5/7.5MA	0.5/1MO		25P	5P	DUA	INB		2N5521	0
2N5522X2	NJD	B51	40V	7.5MA	200C	0.5WF	0.7/4V	.5/7.5MA	0.5/1MO		25P	5P	DUA	INB		2N5521	0
2N5523	NJD	B51	40V	7.5MA	200C	0.5WF	0.7/4V	.5/7.5MA	0.5/1MO		25P	5P	DUA	INB		2N5522	0
2N5523/D	NJD	B73	40V	7.5MA	200C	0.5WF	0.7/4V	.5/7.5MA	0.5/1MO		25P	5P	DUA	INB		2N5522	0
2N5523/W	NJD	B74	40V	7.5MA	200C	0.5WF	0.7/4V	.5/7.5MA	0.5/1MO		25P	5P	DUA	INB		2N5522	0
2N5523X2	NJD	B51	40V	7.5MA	200C	0.5WF	0.7/4V	.5/7.5MA	0.5/1MO		25P	5P	DUA	INB		2N5522	0
2N5524	NJD	B51	40V	7.5MA	200C	0.5WF	0.7/4V	.5/7.5MA	0.5/1MO		25P	5P	DUA	INB		2N5523	0

TYPE NO	CONS TRUC TION	CASE & LEAD	V DS MAX	I D MAX	T J MAX	P TOT MAX	VP OR VT	IDSS OR IDOM	G MO	R DS MAX	C ISS MAX	C RSS MAX	USE	SUPP LIER	EUR SUB	USA SUB	ISS
2N5524/D	NJD	B73	40V	7.5MA	200C	0.5WF	0.7/4V	.5/7.5MA	0.5/1MO		25P	5P	DUA	INB		2N5523	0
2N5524/W	NJD	B74	40V	7.5MA	200C	0.5WF	0.7/4V	.5/7.5MA	0.5/1MO		25P	5P	DUA	INB		2N5523	0
2N5524X2	NJD	B51	40V	7.5MA	200C	0.5WF	0.7/4V	.5/7.5MA	0.5/1MO		25P	5P	DUA	INB		2N5523	0
2N5543	NJD	B22	300V	10MA	200C	0.8WF	2/15V	2/10MA	0.75/3MO	2K	10P	2P	ALH	CRY		2N6449	0
2N5544	NJD	B22	200V	10MA	200C	0.8WF	2/15V	2/10MA	0.75/3MO	2K	10P	2P	ALH	CRY		2N6450	0
2N5545	NJD	B20	50V	8MA	175C	.25WF	0.5/4.5V	0.5/8MA	1.5/6MO		6P	2P	DUA	SIU	BFQ10	2N5515	0
2N5545JAN	NJD	B51	50V	8MA	200C	.25WF	4.5MXV	8MXMA	1.5/6MO		6P		DUA	SIU	BFQ25	2N5196	0
2N5545X2	NJD	B20	50V	8MA	175C	.25WF	0.5/4.5V	0.5/8MA	1.5/6MO		6P	2P	DUA	SIU	BFQ10	2N5515	0
2N5546	NJD	B20	50V	8MA	175C	.25WF	0.5/4.5V	0.5/8MA	1.5/6MO		6P	2P	DUA	SIU	BFQ10	2N5515	0
2N5546JAN	NJD	B51	50V	8MA	200C	.25WF	4.5MX5	8MXMA	1.5/6MO		6P		DUA	SIU	BFQ25	2N5196	0
2N5546X2	NJD	B20	50V	8MA	175C	.25WF	0.5/4.5V	0.5/8MA	1.5/6MO		6P	2P	DUA	SIU	BFQ10	2N5515	0
2N5547	NJD	B20	50V	8MA	175C	.25WF	0.5/4.5V	0.5/8MA	1.5/6MO		6P	2P	DUA	SIU	BFQ10	2N5515	0
2N5547CHP	NJD	B73	50V	8MA	200C		0.5/4.5V	0.5/8MA	1.5/6MO		6P	2P	DUA	SIU			0
2N5547JAN	NJD	B51	50V	8MA	200C	.25WF	4.5MXV	8MXMA	1.5/6MO		6P		DUA	SIU	BFQ25	2N5196	0
2N5547X2	NJD	B20	50V	8MA	175C	.25WF	0.5/4.5V	0.5/8MA	1.5/6MO		6P	2P	DUA	SIU	BFQ10	2N5515	0
2N5548	PME	B49	25V	125MA	150C	.36WF	5MXV	120MXMA	3.5/6.5MO		10P		ALC	TIB		3N161	0
2N5549	NJD	B15	40V		200C	.36WF			6/15MO	100R	8P		RLS	TIB	BSV80	2N4093	0
2N5555	NJD	B3	25V	100MA	125C	0.3WF	5MXV	15MNMA		150R			RLS	INB	BSV80	2N4092	0
2N5555/D	NJD	B73	25V	100MA	125C	0.3WF	5MXV	15MNMA		150R			RLS	INB	BSV80	2N4092	0
2N5555/W	NJD	B74	25V	100MA	125C	0.3WF	5MXV	15MNMA		150R			RLS	INB	BSV80	2N4092	0
2N5556	NJD	B15	30V	10MA	175C	0.3WF	0.2/4V	.5/2.5MA	1.5/6.5MO		6P	3P	ALN	SIU	BFW12	2N3685A	0
2N5556CHP	NJD	B73	40V	30MA	200C		0.5/3V	5/30MA	7.5/12MO		12P	3P	DUA	SIU			0
2N5557	NJD	B15	30V	10MA	175C	0.3WF	0.8/5V	2/5MA	1.5/6.5MO		6P	3P	ALN	SIU	BFW56	2N4869A	0
2N5558	NJD	B15	30V	10MA	175C	0.3WF	1.5/6V	4/10MA	1.5/6.5MO		6P	3P	ALN	SIU	BFW11	2N4869A	0
2N5561	NJD	B51	50V	10MA	200C	.25WF	3MXV	10MXMA	2/3MO		7P		DUA	GIU	BFQ10	2N4084	0

TYPE NO	CONS TRUC TION	CASE & LEAD	V DS MAX	I D MAX	T J MAX	P TOT MAX	VP OR VT	IDSS OR IDOM	G MO	R DS MAX	C ISS MAX	C RSS MAX	USE	SUPP LIER	EUR SUB	USA SUB	ISS
2N5562	NJD	B51	50V	10MA	200C	.25WF	3MXV	10MXMA	2/3MO		7P		DUA	GIU	BFQ10	2N5561	0
2N5563	NJD	B51	50V	10MA	200C	.25WF	3MXV	10MXMA	2/3MO		7P		DUA	GIU	BFQ10	2N5561	0
2N5564	NJD	B20	40V	30MA	150C	.33WF	0.5/3V	5/30MA	7.5/12MO	100R	12P	3P	DUA	SIU		2N5911	0
2N5564X2	NJD	B20	40V	30MA	150C	.33WF	0.5/3V	5/30MA	7.5/12MO	100R	12P	3P	DUA	SIU		2N5911	0
2N5565	NJD	B20	40V	30MA	150C	.33WF	0.5/3V	5/30MA	7.5/12MO		12P	3P	DUA	SIU		2N5591	0
2N5565X2	NJD	B20	40V	30MA	150C	.33WF	0.5/3V	5/30MA	7.5/12MO	100R	12P	3P	DUA	SIU		2N5911	0
2N5566	NJD	B20	40V	30MA	150C	.33WF	0.5/3V	5/30MA	7.5/12MO	100R	12P	3P	DUA	SIU		2N5911	0
2N5566X2	NJD	B20	40V	30MA	150C	.33WF	0.5/3V	5/30MA	7.5/12MO	100R	12P	3P	DUA	SIU		2N5911	0
2N5592	NJD	B15	50V	10MA	200C	0.3WF	5MXV	1/10MA	2/7MO		20P		ALN	OBS	BFW56	2N3684A	0
2N5593	NJD	B15	50V	10MA	200C	0.3WF	5MXV	1/10MA	2/7MO		20P		ALN	OBS	BFW56	2N3684A	0
2N5594	NJD	B15	50V	10MA	200C	0.3WF	5MXV	1/10MA	2/7MO		20P		ALN	OBS	BFW56	2N3684A	0
2N5638	NJD	B3	30V	250MA	135C	.31WF	12MXV	50MNMA		30R	10P	4P	RLS	INB	BSV78	2N4391	0
2N5639	NJD	B3	30V	250MA	135C	.31WF	8MXV	25MNMA		60R	10P	4P	RLS	INB	BSV80	2N4392	0
2N5640	NJD	B3	30V	250MA	135C	.31WF	6MXV	5MNMA		100R	10P	4P	RLS	INB	BSV80	2N4393	0
2N5647	NJD	B15	50V	2MA	150C	0.3WF	0.6/1.8V	0.3/.6MA	0.3/.65MO		3P	0P9	ALN	SIU			0
2N5648	NJD	B15	50V	2MA	150C	0.3WF	0.8/2.4V	0.5/1MA	0.4/0.8MO		3P	0P9	ALN	SIU			0
2N5649	NJD	B15	50V	2MA	150C	0.3WF	1/3V	.8/1.6MA	0.45/.9MO		3P	0P9	ALN	SIU			0
2N5653	NJD	B3	30V	200MA	150C	.31WF	12MXV	40MNMA		50R	10P	3P5	RLS	TDY	BSV79	2N5653	0
2N5654	NJD	B3	30V	150MA	150C	.31WF	12MXV	15MNMA		100R	10P	3P5	RLS	TDY	BSV80	2N4093	0
2N5668	NJD	B3	25V	5MA	150C	.31WF	0.2/4V	1/5MA	1MNMO		7P	3P	RLG	TDY	BC264A	2N5457	0
2N5669	NJD	B3	25V	10MA	150C	.31WF	1/6V	4/10MA	1.6MNMO		7P	3P	RLG	TDY	BC264C	2N5458	0
2N5670	NJD	B3	25V	20MA	150C	.31WF	2/8V	8/20MA	2.5MNMO		7P	3P	RLG	TDY	BFW61	2N5486	0
2N5716	NJD	B3	40V	4MA	150C	.35WF	0.2/3V	0.25MXMA	0.2/1MO		5P	1P	ALN	MOB	BF800	2N3687	0
2N5717	NJD	B3	40V	4MA	150C	0.35W	0.5/5V	0.2/1MA	0.4/1.6MO		5P	1P	ALN	MOB	BF800	2N3686	0
2N5718	NJD	B3	40V	4MA	150C	0.35W	1/8V	0.8/4MA	0.8/4MO		5P	1P	ALN	MOB	BC264A	2N4868	0

TYPE NO	CONS TRUC TION	CASE & LEAD	V DS MAX	I D MAX	T J MAX	P TOT MAX	VP OR VT	IDSS OR IDOM	G MO	R DS MAX	C ISS MAX	C RSS MAX	USE	SUPP LIER	EUR SUB	USA SUB	ISS
2N5797	PJD	B63	40V	2MA	150C	.35WF	0.5/4V	.02/.1MA	.06/.23MO		5P	1P	ALG	MOB		2N2606	0
2N5798	PJD	B63	40V	2MA	150C	.35WF	0.8/6V	.08/.4MA	0.1/.4MO		5P	1P	ALG	MOB		2N2606	0
2N5799	PJD	B63	40V	2MA	150C	.35WF	1.2/8V	0.25/1MA	0.16/.5MO		5P	1P	ALG	MOB		2N2843	0
2N5800	PJD	B63	40V	2MA	150C	.35WF	2/9V	0.7/2MA	0.25/.7MO		5P	1P	ALG	MOB		2N2607	0
2N5801	NJD	B3	40V	15MA	150C	.35WF	8MXV	2/15MA	0.16MNMO				RMS	MOB		2N4393	0
2N5802	NJD	B3	40V	40MA	150C	.35WF	8MXV	10/40MA	0.16MNMO				RMS	MOB		2N4392	0
2N5803	NJD	B3	40V	80MA	150C	.35WF	8MXV	30/80MA	0.16MNMO				RMS	MOB		2N4391	0
2N5902	NJD	B43	40V	0.5MA	150C	0.5WF	0.6/4.5V	.03/.5MA	.07/.25MO		3P	1P5	DUA	INB		2N5906	0
2N5902/D	NJD	B73	40V	0.5MA	150C	0.5WF	0.6/4.5V	.03/.5MA	.07/.25MO		3P	1P5	DUA	INB		2N5906	0
2N5902/W	NJD	B74	40V	0.5MA	150C	0.5WF	0.6/4.5V	.03/.5MA	.07/.25MO		3P	1P5	DUA	INB		2N5906	0
2N5903	NJD	B43	40V	0.5MA	150C	0.5WF	0.6/4.5V	.03/.5MA	.07/.25MO		3P	1P5	DUA	INB		2N5907	0
2N5903/D	NJD	B73	40V	0.5MA	150C	0.5WF	0.6/4.5V	.03/.5MA	.07/.25MO		3P	1P5	DUA	INB		2N5907	0
2N5903/W	NJD	B74	40V	0.5MA	150C	0.5WF	0.6/4.5V	.03/.5MA	.07/.25MO		3P	1P5	DUA	INB		2N5907	0
2N5904	NJD	B43	40V	0.5MA	150C	0.5WF	0.6/4.5V	.03/.5MA	.07/.25MO		3P	1P5	DUA	INB		2N5908	0
2N5904/D	NJD	B73	40V	0.5MA	150C	0.5WF	0.6/4.5V	.03/.5MA	.07/.25MO		3P	1P5	DUA	INB		2N5908	0
2N5904/W	NJD	B74	40V	0.5MA	150C	0.5WF	0.6/4.5V	.03/.5MA	.07/.25MO		3P	1P5	DUA	INB		2N5908	0
2N5905	NJD	B43	40V	0.5MA	150C	0.5WF	0.6/4.5V	.03/.5MA	.07/.25MO		3P	1P5	DUA	INB		2N5909	0
2N5905/D	NJD	B73	40V	0.5MA	150C	0.5WF	0.6/4.5V	.03/.5MA	.07/.25MO		3P	1P5	DUA	INB		2N5909	0
2N5905/W	NJD	B74	40V	0.5MA	150C	0.5WF	0.6/4.5V	.03/.5MA	.07/.25MO		3P	1P5	DUA	INB		2N5909	0
2N5905CHP	NJD	B73	40V	1MA	200C		0.6/4.5V	.03/.5MA	.07/.25MO		3P	1P5	DUA	SIU			0
2N5906	NJD	B43	40V	1MA	150C	.37WF	4.5MXV	0.5MXMA	.07/.25MO		3P		DUA	INB		2N5902	0
2N5906/D	NJD	B73	40V	0.5MA	150C	0.5WF	0.6/4.5V	.03/.5MA	.07/.25MO		3P	1P5	DUA	INB		N	0
2N5906/W	NJD	B74	40V	0.5MA	150C	0.5WF	0.6/4.5V	.03/.5MA	.07/.25MO		3P	1P5	DUA	INB			0
2N5907	NJD	B43	40V	0.5MA	150C	0.5WF	0.6/4.5V	.03/.5MA	.07/.25MO		3P	1P5	DUA	INB		2N5906	0
2N5907/D	NJD	B73	40V	0.5MA	150C	0.5WF	0.6/4.5V	.03/.5MA	.07/.25MO		3P	1P5	DUA	INB		2N5906	0

TYPE NO	CONS TRUC TION	CASE & LEAD	V DS MAX	I D MAX	T J MAX	P TOT MAX	VP OR VT	IDSS OR IDOM	G MO	R DS MAX	C ISS MAX	C RSS MAX	USE	SUPP LIER	EUR SUB	USA SUB	ISS
2N5907/W	NJD	B74	40V	0.5MA	150C	0.5WF	0.6/4.5V	.03/.5MA	.07/.25MO		3P	1P5	DUA	INB		2N5906	0
2N5908	NJD	B43	40V	0.5MA	150C	0.5WF	0.6/4.5V	.03/.5MA	.07/.25MO		3P	1P5	DUA	INB		2N5907	0
2N5908/D	NJD	B73	40V	0.5MA	150C	0.5WF	0.6/4.5V	.03/.5MA	.07/.25MO		3P	1P5	DUA	INB		2N5907	0
2N5908/W	NJD	B74	40V	0.5MA	150C	0.5WF	0.6/4.5V	.03/.5MA	.07/.25MO		3P	1P5	DUA	INB		2N5907	0
2N5909	NJD	B43	40V	0.5MA	150C	0.5WF	0.6/4.5V	.03/.5MA	.07/.25MO		3P	1P5	DUA	INB		2N5908	0
2N5909/D	NJD	B73	40V	0.5MA	150C	0.5WF	0.6/4.5V	.03/.5MA	.07/.25MO		3P	1P5	DUA	INB		2N5908	0
2N5909/W	NJD	B74	40V	0.5MA	150C	0.5WF	0.6/4.5V	.03/.5MA	.07/.25MO		3P	1P5	DUA	INB		2N5908	0
2N5911	NJD	B43	25V	40MA	150C	0.5WF	1/5V	7/40MA	5/10MO		5P	1P2	DUA	INB		2N5564	0
2N5911/D	NJD	B73	25V	40MA	150C	0.5WF	1/5V	7/40MA	5/10MO		5P	1P2	DUA	INB		2N5564	0
2N5911/W	NJD	B74	25V	40MA	150C	0.5WF	1/5V	7/40MA	5/10MO		5P	1P2	DUA	INB		2N5564	0
2N5912	NJD	B43	25V	40MA	150C	0.5WF	1/5V	7/40MA	5/10MO		5P	1P2	DUA	INB		2N5911	0
2N5912/D	NJD	B73	25V	40MA	150C	0.5WF	1/5V	7/40MA	5/10MO		5P	1P2	DUA	INB		2N5911	0
2N5912/W	NJD	B74	25V	40MA	150C	0.5WF	1/5V	7/40MA	5/10MO		5P	1P2	DUA	INB		2N5911	0
2N5949	NJD	B10	30V	18MA	150C	.35WF	3/7V	12/18MA	3.5/7.5MO	200R	6P	2P	RLS	TDY	BF244	2N3824	0
2N5950	NJD	B10	30V	18MA	150C	.35WF	2.5/6V	10/15MA	3.5/7.5MO	210R	6P	2P	ALG	TDY	BF244	2N3824	0
2N5951	NJD	B10	30V	18MA	150C	.35WF	2/5V	7/13MA	3.5/6.5MO	250R	6P	2P	ALG	TDY	BC264D	2N3824	0
2N5952	NJD	B10	30V	8MA	150C	.35WF	1.3/3.5V	4/8MA	2/6.5MO	300R	6P	2P	ALG	TDY	BC264C	2N4303	0
2N5953	NJD	B10	30V	8MA	150C	.35WF	0.8/3V	2.5/5MA	2/6.5MO	375R	6P	2P	ALG	TDY	BC264B	2N4302	0
2N6449	NJD	B22	300V	10MA	200C	0.8WF	2/15V	2/10MA	0.5/3MO		10P	5P	ALH	CRY		2N5543	0
2N6450	NJD	B22	200V	10MA	200C	0.8WF	2/15V	2/10MA	0.5/3MO		10P	5P	ALH	CRY		2N5544	0
2N6451	NJD	B15	20V	20MA	200C	.36WF	3.5MXV	20MXMA	15/30MO		25P		ALN	TIB	BF817	2N6452	0
2N6452	NJD	B15	25V	20MA	200C	.36WF	3.5MXV	20MXMA	15/30MO		25P		ALN	TIB	BF817	2N6451	0
2N6453	NJD	B15	20V	50MA	200C	.36WF	5MXV	50MXMA	20/40MO		25P		ALN	TIB	BF817	2N6454	0
2N6454	NJD	B15	25V	50MA	200C	.36WF	5MXV	50MXMA	20/40MO		25P		ALN	TIB	BF817	2N6453	0
2N6483	NJD	B51	50V	7.5MA	150C	0.5WF	0.7/4V	.5/7.5MA	1/4MO		20P	3P5	DUA	INB		2N5196	0

TYPE NO	CONS TRUC TION	CASE & LEAD	V DS MAX	I D MAX	T J MAX	P TOT MAX	VP OR VT	IDSS OR IDOM	G MO	R DS MAX	C ISS MAX	C RSS MAX	USE	SUPP LIER	EUR SUB	USA SUB	ISS
2N6484	NJD	B51	50V	7.5MA	150C	0.5WF	0.7/4V	.5/7.5MA	1/4MO		20P	3P5	DUA	INB		2N5196	0
2N6485	NJD	B51	50V	7.5MA	150C	0.5WF	0.7/4V	.5/7.5MA	1/4MO		20P	3P5	DUA	INB		2N5196	0
2N6550	NJD	B47	20V	300MA	200C	0.4WF	0.3/3V	10MNMA	25MNMO		35P	20P	ALN	CRY	BF817	2N6453	0
2N6568	NJD	B6	30V	1A	200C	0.4WF	10MXV	500MNMA		2R5	60P		RMS	CRY			0
2SJ11	PJD	B15	20V	5MA	125C	0.1WF	5MXV	0.9MXMA	0.1/0.6MO		2P		ALG	TOB		2N2606	0
2SJ12	PJD	B15	20V	5MA	125C	0.1WF	5MXV	0.9MXMA	0.1/0.6MO		2P		ALG	TOB		2N2606	0
2SJ13	PJD	B12	12V	10MA	150C	0.1WF	4.5MXV	5/10MA					ALG	OBS	BF320C	2N3331	0
2SJ15	PJD	B6	20V	10MA	175C	0.2WF	6MXV	1.5MXMA	0.2/3MO		6P		ALG	FUJ	BF320A	2N2607	0
2SJ16	PJD	B6	20V	10MA	175C	0.2WF	6MXV	1.5MXMA	0.2/3MO		6P		ALG	FUJ	BF320A	2N2607	0
2SJ22	PJD	B13	80V	1MA	85C	50MWF		0.7MXMA	200MNMO				ALG				0
2SK11	NJD	B15	20V	10MA	150C	0.1WF	0.5/6V	.3/6.5MA	0.7/3.2MO		5P		ALG	TOB	BFW12	2N5457	0
2SK11/O	NJD	B15	20V	10MA	150C	0.1WF	0.8/3.5V	.8/2.5MA	1/3MO		5P		ALG	TOB	BFW12	2N4302	0
2SK11/R	NJD	B15	20V	10MA	150C	0.1WF	0.5/2V	0.3/1MA	0.7/2.3MO		5P		ALG	TOB	BFW13	2N3821	0
2SK11/Y	NJD	B15	20V	10MA			1.4/6V	2/6.5MA	1.3/3.2MO		5P		ALG	TOB	BC264B	2N3684	0
2SK12	NJD	B15	20V	10MA	150C	0.1WF	.65/4.5V	0.45/5MA	0.8/3.2MO		5P		ALN	TOB	BF347	2N5457	0
2SK12/G	NJD	B15	20V	10MA	150C	0.1WF	1.7/4.5V	2.5/5MA	1.6/3.2MO		5P		ALN	TOB	BC264A	2N3684	0
2SK12/O	NJD	B15	20V	10MA	150C	0.1WF	0.9/2.2V	.8/1.6MA	1/2.3MO		5P		ALN	TOB	BFW13	2N3821	0
2SK12/R	NJD	B15	20V	10MA	150C	0.1WF	.65/1.6V	.45/.9MA	0.8/1.9MO		5V		ALN	TOB	BFW13	2N3686	0
2SK12/Y	NJD	B15	20V	10MA	150C	0.1WF	1.2/3V	1.4/3MA	1.3/3MO		5P		ALN	TOB	BF808	2N3685	0
2SK13	NJD	B15	12V	10MA	150C	0.1WF	.65/4.5V	0.45/5MA	0.8/3.5MO		3P5		ALN	OBS	BFW12	2N5457	0
2SK13/G	NJD	B15	12V	10MA	150C	0.1WF	1.7/4.5V	2.5/5MA	1.6/3.2MO		3P5		ALN	OBS	BC264B	2N3684	0
2SK13/O	NJD	B15	12V	10MA	150C	0.1WF	0.9/2.2V	.8/1.6MA	1/2.3MO		3P5		ALN	OBS	BF808	2N3685	0
2SK13/R	NJD	B15	12V	10MA	150C	0.1WF	.65/1.6V	.45/.9MA	0.8/1.9MO		3P5		ALN	OBS	BF800	2N3686	0
2SK13/Y	NJD	B15	12V	10MA	150C	0.1WF	1.2/3V	1.4/3MA	1.3/3MO		3P5		ALN	OBS	BC264A	2N3685A	0
2SK15	NJD	B15	20V	10MA	150C	0.1WF	0.65/5V	0.45/5MA	0.8/3.2MO		5P		ALN	TOB	BF808	2N4868	0

TYPE NO	CONS TRUC TION	CASE & LEAD	V DS MAX	I D MAX	T J MAX	P TOT MAX	VP OR VT	IDSS OR IDOM	G MO	R DS MAX	C ISS MAX	C RSS MAX	USE	SUPP LIER	EUR SUB	USA SUB	ISS
2SK15/G	NJD	B15	20V	10MA	150C	0.1WF	1.8/5V	2.5/5MA	1.6/3.2MO		5P		ALN	TOB	BC264A	2N3684	0
2SK15/0	NJD	B15	20V	10MA	150C	0.1WF	0.9/2.5V	.8/1.6MA	1/2.3MO		5P		ALN	TOB	BFW13	2N3686A	0
2SK15/O	NJD	B15	20V	10MA	150C	0.1WF	0.9/2.5V	.8/1.6MA	1/2.3MO		5P		ALN	TOB	BFW13	2N3686A	0
2SK15/R	NJD	B15	20V	10MA	150C	0.1WF	.65/1.8V	.45/.9MA	0.8/1.9MO		5P		ALN	TOB	BF808	2N3686A	0
2SK15/Y	NJD	B15	20V	10MA	150C	0.1WF	1.3/3.5V	1.4/3MA	1.3/3MO		5P		ALN	TOB	BC264A	2N3685A	0
2SK16H	NJD	B15	20V	10MA	150C	0.1WF	3MXV	0.5/7MA	1/6MO		9P		ALG	HIJ	BFW12	2N5458	0
2SK16HA	NJD	B15	20V	10MA	150C	0.1WF	3MXV	0.5/7MA	1/2.5MO		9P		ALG	HIJ	BFW12	2N4032	0
2SK16HB	NJD	B15	20V	10MA	150C	0.1WF	3MXV	0.5/7MA	2/4MO		9P		ALG	HIJ	BFW12	2W5458	0
2SK16HC	NJD	B15	20V	10MA	150C	0.1WF	3MXV	0.5/7MA	3/6MO		9P		ALG	HIJ	BFW61	2N3822	0
2SK17	NJD	B3	20V	10MA	125	0.1WF	6MXV	6.5MXMA	0.7MNMO		6P		ALN	OBS	BF808	2N3685A	0
2SK18	NJD		20V	3MA	150C	0.2WF	3.5MXV	3MXMA	0.8/3MO		6P		FVG	TOS	BC264A	2N3685A	0
2SK18A	NJD		20V	3MA	150C	0.2WF	3.5MXV	3MXMA	0.8/3MO		6P		FVG	TOS	BC264A	2N3685A	0
2SK19	NJD	B31	8V	25MA	125C	0.2WF	1.2/7V	3/24MA	7TPMO		0P65		FVG	TOS	BF256	2N5245	0
2SK19/B	NJD	B31	8V	25MA	125C	0.2WF	1.2/7V	12/24MA	7TPMO		0	P65	FVG	TOS	BF256C	2N5247	0
2SK19/G	NJD	B31	8V	25MA	125C	0.2WF	1.2/7V	6/14MA	7TPMO		0	P65	FVG	TOS	BF256B	2N5245	0
2SK19/Y	NJD	B3	18V	25MA	125C	0.2WF	1.2/7V	3/7MA	7TPMO		0	P65	FVG	TOS	BF256A	2N5246	0
2SK23	NJD	B1	18V	20MA	150C	.15WF	4.9MXV	16MXMA	2.7MNMO		5P		ALG	SOY	BFW61	2N5459	0
2SK23A	NJD	B1	18V	20MA	150C	.15WF	4.9MXV	15/20MA	4MNMO		5P		ALG		BFW61	2N5459	0
2SK24C	NJD	OBS	40V	2MA	125C	0.1WF	6MXV	1.5MXMA	1.5/12MO		8P		ALG	OBS	BFW13	2N3821	0
2SK24D	NJD	OBS	40V	5MA	125C	0.1WF	6MXV	3MXMA	1.5/12MO		8P		ALG	OBS	BF347	2N3685	0
2SK24E	NJD	OBS	40V	10MA	125C	0.1WF	6MXV	6MXMA	1.5/12MO		8P		ALG	OBS	BFW12	2N5457	0
2SK24F	NJD	OBS	40V	12MA	125C	0.1WF	6MXV	12MXMA	1.5/12MO		8P		ALG	OBS	BC264	2N5459	0
2SK24G	NJD	OBS	40V	25MA	125C	0.1WF	6MXV	24MXMA	1.5/12MO		8P		ALG	OBS	BF244	2N3819	0
2SK30	NJD	B1	50V	10MA	125C	0.1WF	0.4/5V	.3/6.5MA	1.5MNMO		16P	5P	ALG	TOB	BF347	2N5458	0
2SK30/G	NJD	B1	50V	10MA	125C	0.1WF	0.4/5V	2.6/7MA	1.5MNMO		16P	5P	ALG	TOB	BC264A	2N5458	0

TYPE NO	CONS TRUC TION	CASE & LEAD	V DS MAX	I D MAX	T J MAX	P TOT MAX	VP OR VT	IDSS OR IDOM	G MO	R DS MAX	C ISS MAX	C RSS MAX	USE	SUPP LIER	EUR SUB	USA SUB	ISS
2SK30/O	NJD	B1	50V	10MA	125C	0.1WF	0.4/5V	.6/1.4MA	1.5MNMO		16P	5P	ALG	TOB	BF347	2N3821	0
2SK30/R	NJD	B1	50V	10MA	125C	0.1WF	0.4/5V	.3/.75MA	1.5MNMO		16P	5P	ALG	TOB	BF347	2N3686	0
2SK30/Y	NJD	B1	50V	10MA	125C	0.1WF	0.4/5V	1.2/3MA	1.5MNMO		16P	5P	ALG	TOB	BFW12	2N3685	0
2SK30A	NJD	B1	50V	10MA	125C	0.1WF	0.4/5V	.3/6.5MA	1.2MNMO		16P	5P	ALN	TOS	BF347	2N5458	0
2SK30A/G	NJD	B1	50V	10MA	125C	0.1WF	0.4/5V	2.6/6MA	1.2MNMO		16P	5P	ALN	TOS	BC264A	2N5458	0
2SK30A/O	NJD	B1	50V	10MA	125C	0.1WF	0.4/5V	.6/1.4MA	1.2MNMO		16P	5P	ALN	TOS	BF347	2N3821	0
2SK30A/R	NJD	B1	50V	10MA	125C	0.1WF	0.4/5V	.3/.75MA	1.2MNMO		16P	5P	ALN	TOS	BF347	2N3686	0
2SK30A/Y	NJD	B1	50V	10MA	125C	0.1WF	0.4/5V	1.2/3MA	1.2MN		16P	5P	ALN	TOS	BFW12	2N3685	0
2SK32	NJD	B1	35V	20MA	150C	0.3WF	4MXV	10MXMA	3MNMO		10P		ALG	NIP	BC264	2N3822	0
2SK33	NJD	B3	20V	20MA	125C	.15WF	8MXV	20MXMA	7TPMO		3P5		ALN	MIT	BF810	2N5397	0
2SK34	NJD	B1	30V	20MA	125C	.15WF	6MXV	6MXMA	3TPMO		6P		ALN	MIT	BFW56	2N3684A	0
2SK35	NMD	B1	20V	20MA	150C	0.2WF	1.6MXV	16MXMA	6.3/15MO		15P		ALG	SOY	BFR29	3N128	0
2SK37	NJD	B69	15V	20MA	120C	0.1WF	4MXV	6MXMA	1.5MNMO		4P		FVG	NIP	BFW12	2N5246	0
2SK38	NMD	B1	30V	10MA	125C	.15WF	6MXV	0.5/5MA	0.8MNMO				ALG	OBS	BFX63	2N3631	0
2SK39A	NJD	B69	20V	2MA	120C	0.2WF		0.6MNMA	4MNMO				ALG	SOY	BF347	2N4302	0
2SK40	NJD		50V	10MA	125C	0.1WF	5MXV	.3/6.5MA	1MNMO				ALG	HIJ	BF808	2N4302	0
2SK40A	NJD		50V	10MA	125C	0.1WF	5MXV	0.3/.8MA	1MNMO				ALG	HIJ	BFW13	2N3686	0
2SK40B	NJD		50V	10MA	125C	0.1WF	5MXV	.6/1.4MA	1MNMO				ALG	HIJ	BFW13	2N4867	0
2SK40C	NJD		50V	10MA	125C	0.1WF	5MXV	1.2/3MA	1MNMO				ALG	HIJ	BC264A	2N5457	0
2SK40D	NJD		50V	10MA	125C	0.1WF	5MXV	2.6/7MA	1MNMO				ALG	HIJ	BFW12	2N5458	0
2SK41	NJD	B3	40V	15MA	125C	0.1WF	5MXV	12MXMA	7TPMO				ALG		BC264D	2N5459	0
2SK41C	NJD	B3	40V	15MA	125C	0.1WF	1.2MXV	1.5MXMA	7TPMO				ALG		BFW13	2N4867	0
2SK41D	NJD	B3	40V	15MA	125C	0.1WF	1.9MXV	3MXMA	7TPMO				ALG		BC264A	2N3685	0
2SK41E	NJD	B3	40V	15MA	125C	0.1WF	3MXV	6MXMA	7TPMO						BF256A	2N5246	0
2SK41F	NJD	B3	40V	15MA	125C	0.1WF	5MXV	12MXMA	7TPMO				ALG		BF810	2N3822	0

TYPE NO	CONS TRUC TION	CASE & LEAD	V DS MAX	I D MAX	T J MAX	P TOT MAX	VP OR VT	IDSS OR IDOM	G MO	R DS MAX	C ISS MAX	C RSS MAX	USE	SUPP LIER	EUR SUB	USA SUB	ISS
2SK42	NJD	B10	12V	8MA	120C	.05WF		3/5MA	4.5MNMO				ALG		BC264B	2N3822	0
2SK43	NJD	B2	30V	20MA	125C	0.3WF	1.1MXV	7/10MA	13MNMO				ALG		BF817	2N6451	0
2SK44	NJD	B1	10V	1MA	125C	0.1WF	4MXV	0.2/.4MA	2MNMO				ALG	TOS	BFW13	2N3687	0
2SK46	NJD	B1	30V	12MA	125C	.15WF	6MXV	0.3/12MA	3MNMO				ALG	OBS	BC264	2N5459	0
2SK48	NJD	B12	20V	10MA	150C	0.1WF	35/2.3V	0.3/3MA	1/5MA		8P		ALN	TOB	BF347	2N3685A	0
2SK49	NJD	B2	20V	10MA	80C	70MWF	2.5MXV	2TPMA	0.5/6MO		5P		FVG	NIP	BFW61	2N5484	0
2SK50	NJD	B1	10V	2MA	80C			1MXMA					FVG	MAT	BF800	2N3686A	0
2SK54	NJD	B2	15V	5MA	125C	.13WF	5MXV	3/5MA	4.5MNMO				ALG	HIJ	BF805	2N3822	0
2SK55	NJD	B2	15V	5MA	125C	.13WF	5MXV	3/5MA	5.5MNMO				ALG	HIJ	BF805	2N3822	0
2SK56	NJD	B3	10V	10MA	120C	.12WF		8/10MA	4MNMO				ALG	MAT	BF244	2N4303	0
2SK57	NJD	B59	20V	10MA	120C	.11WF	4MXV	4/6MA	1.5MNMO				ALG	NIP	BFR30		0
2SK61	NJD		20V	10MA	125C	0.2WF	4MXV	10MXMA	9TPMO				FVG	TOS	BF256B	2N5105	0
2SK65	NJD	B1	20V	1MA	80C			0.8MXMA	0.3MNMO				ALN	MAT		2N3687A	0
2SK66	NJD	B1	20V	10MA	125C	0.1WF		6.5MXMA	1.2MNMO		8P		ALG	MAT	BF808	2N3685A	0
2SK68	NJD	B3	50V	12MA	125C	.25WF	1.5MXV	12MXMA	4MNMO		13P		ALG	NIP	BF810	2N3822	0
2SK68A	NJD	B3	50V	12MA	125C	.25WF	1.5MXV	12MXMA	4MNMO		13P		ALN	NIP	BF810	2N3822	0
2SK72	NJD	B51	20V	7MA	150C	0.2WF	0.5MXV	6.5MXMA	1.5/6.5MO				DUA	TOS	BFQ10	2N5452	0
3G2	NJD	B15	50V	4MA	200C	0.3WF	3MXV	4MXMA	1MNMO		7P5		ALG	OBS	BFW12	2N3821	0
3N89	PJD	B57	30V	50MA	200C	0.3WF	4MXV	2.5MXMA	.45/1.3			3P	ALG	SIU			0
3N96	PJD	B46	30V	50MA	175C	0.3WF	4MXV	2.5MXMA	.45/1.3MO		4P		DUA	OBS		2N5510	0
3N97	PJD	B46	30V	50MA	175C	0.3WF	4MXV	2.5MXMA	.45/1.3MO		4P		DUA	OBS		2N5510	0
3N98	NMD	B56	32V	8MA	125C	.15WF	6MXV	3.5/8MA	1/3MO		7P		ALG	OBS		2N3631	0
3N99	NMD	B56	32V	15MA	85C	.15WF	6MXV	5/10.5MA	1/4.5MO	900R	7P		ALG	OBS		2N3631	0
3N124	NJD	B57	50V	20MA	175C	0.3WF	2.5MXV	0.2/2MA	0.5/2MO		5P	OP5	ALG	MOB			0
3N125	NJD	B57	50V	20MA	175C	0.3WF	4MXV	1.5/4MA	0.8/2.4MO		5P	OP5	ALG	MOB			0

TYPE NO	CONS TRUC TION	CASE & LEAD	V DS MAX	I D MAX	T J MAX	P TOT MAX	VP OR VT	IDSS OR IDOM	G MO	R DS MAX	C ISS MAX	C RSS MAX	USE	SUPP LIER	EUR SUB	USA SUB	ISS
3N126	NJD	B57	50V	20MA	175C	0.3WF	6.5MXV	3/9MA	1.2/3.6MO		5P	OP5	ALG	MOB			0
3N128	NMD	B54	20V	25MA	175C	.33WF	8MXV	5/25MA	5/12MO	200R	7P	OP28	FVG	RCB	BFR29	3N128	0
3N138	NMD	B54	35V	50MA	175C	.33WF	6MXV	25MXMA	6TPMO	300R	5P	OP25	ALC	RCB	BSV81	3N128	0
3N139	NMD	B54	35V	50MA	175C	.33WF	6MXV	5/25MA	3/6MO		7P	OP3	FVG	RCB	BSV81	3N128	0
3N140	NMD	B66	20V	50MA	175C	0.4WF	4MXV	5/30MA	6/18MO		7P	OP03	FVG	MOB	BF352	3N187	0
3N140X2	NMD	B66	20V	50MA	175C	0.4WF	4MXV	5/30MA	6/18MO		7P	OP03	MPP	OBS	BF352X2	3N187X2	0
3N141	NMD	B66	20V	50MA	175C	0.4WF	4MXV	5/30MA	6/18MO		7P	OP03	FVG	RCB	BF352	3N187	0
3N141X2	NMD	B66	20V	50MA	175C	0.4WF	4MXV	5/30MA	6/18MO		7P	OP03	FVG	RCB	BF352X2	3N187X2	0
3N142	NMD	B54	20V	50MA	175C	.33WF	0.5/8V	5/25MA	5/7.5MO		10P	OP28	FVG	RCB		3N128	0
3N143	NMD	B54	20V	25MA	175C	.33WF	8MXV	5/30MA	5/12MO	200R	7P	OP28	FVG	RCB	BFR29	3N128	0
3N145	PME	B53	30V	20MA	175C	.35WF	6MXV	3MNMA		1K5	1P		ALG	RCB	BSW95	2N4352	0
3N146	PME	B53	30V	20MA	175C	.35WF	6MXV	3MNMA		1K5	1P		ALG	INB	BSW95	3N174	0
3N147	PME	B38	20V	200MA	175C	0.6WF	3/12V	8MNMA		500R	4P	2P	RLS	INB		2N4066	0
3N148	PME	B38	30V	200MA	175C	0.6WF	3/12V	8MNMA		500R	4P	2P	RLS	INB		2N4066	0
3N149	PME	B53	30V	200MA	175C	0.4WF	6MXV	16MNMA		250R	3P		RLS	INB		3N161	0
3N150	PME	B53	30V	200MA	175C	0.4WF	6MXV	16MNMA		250R	3P		RLS	INB		3N161	0
3N151	PME	B42	30V	25MA	150C	.15WF	6MXV	3MXMA	0.5/3MO		12P		DUA	INB		3N208	0
3N152	NMD	B54	20V	50MA	175C	.33WF	8MXV	5/30MA	5/7.5MO	200R	10P	OP28	FVG	RCB		3N128	0
3N153	NMD	B54	20V	50MA	175C	0.4WF	7MXV		10TPMO	300R	8P	OP5	RLS	RCB		3N138	0
3N154	NMD	B54	20V	50MA	175C	.33WF	0.5/8V	10/25MA	5/12MO	200R	7P	OP2	FVG	RCB		3N128	0
3N155	PME	B56	35V	30MA	175C	0.3WF	0.5/3.2V	5MNMA	1/4MO	600R	5P	1P3	ALC	MOB		3N163	0
3N155A	PME	B56	35V	30MA	175C	0.3WF	1.5/3.2V	5MNMA	1/4MO	300R	5P	1P3	ALC	MOB		3N163	0
3N156	PME	B56	35V	30MA	175C	0.3WF	3/5V	5MNMA	1/4MO	600R	5P	1P3	ALC	MOB		3N163	0
3N156A	PME	B56	35V	30MA	175C	0.3WF	3/5V	5MNMA	1/4MO	300R	5P	1P3	ALC	MOB		3N163	0
3N157	PME	B56	35V	30MA	175C	0.3WF	1.5/3.2V	5MNMA	1/4MO		5P	1P3	RLS	MOB		2N4352	0

TYPE NO	CONS TRUC TION	CASE & LEAD	V DS MAX	I D MAX	T J MAX	P TOT MAX	VP OR VT	IDSS OR IDOM	G MO	R DS MAX	C ISS MAX	C RSS MAX	USE	SUPP LIER	EUR SUB	USA SUB	ISS
3N157A	PME	B56	50V	30MA	175C	0.3WF	1.5/3.2V	5MNMA	1/4MO		5P	1P3	RLS	MOB		2N4352	0
3N158	PME	B56	35V	30MA	175C	0.3WF	3/5V	5MNMA	1/4MO	5P		1P3	RLS	MOB		2N4352	0
3N158A	PME	B56	50V	30MA	175C	0.3WF	3/5V	5MNMA	1/4MO		5P	1P3	RLS	MOB		2N4352	0
3N159	NMD	B66	20V	50MA	175C	.33WF	4MXV	5/30MA	7/18MO		7P	0P03	FVG	RCB	BF352	3N202	0
3N160	PME	B53	25V	125MA	200C	.36WF	1.5/5V	40/120MA	3.5/6.5MO		10P	4P	RLS	INB		3N161	0
3N161	PME	B53	25V	125MA	200C	.36WF	1.5/5V	40/120MA	3.5/6.5MO		10P	4P	RMS	INB		2N5548	0
3N162	PME	B53	30V	250MA	150C	0.5WF	5MXV	25MXMA		100R	20P		RLS	GIB		3N167	0
3N163	PME	B53	40V	50MA	150C	.37WF	2/5V	5/30MA	2/4MO	250R	2P5	0P7	RLS	INB		3N156A	0
3N163/D	PME	B73	40V	50MA	150C	.37WF	2/5V	5/30MA	2/4MO	250R	2P5	0P7	RLS	INB		3N156A	0
3N163/W	PME	B74	40V	50MA	150C	.37WF	2/5V	5/30MA	2/4MO	250R	2P5	0P7	RLS	INB		3N156A	0
3N163CHP	PME	B73	40V	50MA	150C		2/5V	5/30MA	2/4MO	250R	2P5	0P7	RLS	SIU			0
3N164	PME	B53	30V	50MA	150C	.38WF	2/5V	3/30MA	1/4MO	300R	2P5	0P7	RLS	SIU			0
3N164CHP	PME	B73	30V	50MA	150C		2/5V	3/30MA	1/4MO	300R	2P5	0P7	RLS	SIU			0
3N165	PME	B38	40V	50MA	150C	.52WF	2/5V	5/30MA	1.5/3MO	300R	3P	0P7	DUA	INB		3N190	0
3N166	PME	B38	40V	50MA	150C	.52WF	2/5V	5/30MA	1.5/3MO	300R	3P	0P7	DUA	INB		3N190	0
3N167	PME	B53	30V	500MA	125C	.22WF	2/6V	200MNMA		20R	35P	12P	RMS	SIU			0
3N168	PME	B53	25V	500MA	125C	.22WF	2/6V	100MNMA		40R	35P	12P	RMS	SIU			0
3N169	NME	B15	25V	30MA	175C	0.3WF	0.5/1.5V	10MNMA	1MNMO	200R	5P	1P3	RLS	INB			0
3N169/D	NME	B73	25V	30MA	175C	0.3WF	0.5/1.5V	10MNMA	1MNMO	200R	5P	1P3	RLS	INB			0
3N169/W	NME	B74	25V	30MA	175C	0.3WF	0.5/1.5V	10MNMA	1MNMO	200R	5P	1P3	RLS	INB			0
3N170	NME	B15	25V	30MA	175C	0.3WF	1/2V	10MNMA	1MNMO	200R	5P	1P3	RLS	INB			0
3N170/D	NME	B73	25V	30MA	175C	0.3WF	1/2V	10MNMA	1MNMO	200R	5P	1P3	RLS	INB			0
3N170/W	NME	B74	25V	30MA	175C	0.3WF	1/2V	10MNMA	1MNMO	200R	5P	1P3	RLS	INB			0
3N171	NME	B15	25V	30MA	175C	0.3WF	1.5/3V	10MNMA	1MNMO	200R	5P	1P3	RLS	INB			0
3N171/D	NME	B73	25V	30MA	175C	0.3WF	1.5/3V	10MNMA	1MNMO	200R	5P	1P3	RLS	INB			0

TYPE NO	CONS TRUC TION	CASE & LEAD	V DS MAX	I D MAX	T J MAX	P TOT MAX	VP OR VT	IDSS OR IDOM	G MO	R DS MAX	C ISS MAX	C RSS MAX	USE	SUPP LIER	EUR SUB	USA SUB	ISS
3N171/W	NME	B74	25V	30MA	175C	0.3WF	1.5/3V	10MNMA	1MNMO	200R	5P	1P3	RLS	INB			0
3N172	PME	B53	40V	50MA	150C	.37WF	2/5V	5/30MA	1.5/4MO	250R	3P5		RLS	INB		3N160	0
3N172/D	PME	B73	40V	50MA	150C	.37WF	2/5V	5/30MA	1.5/4MO	250R			RLS	INB		3N160	0
3N172/W	PME	B74	40V	50MA	150C	.37WF	2/5V	5/30MA	1.5/4MO	250R			RLS	INB		3N160	0
3N173	PME	B53	40V	50MA	200C	.37WF	5MXV	50MXMA	1/4MO	350R	3P5		RLS	INB		3N163	0
3N174	PME	B53	30V	20M	A20C	.36WF	2/6V	3/12MA		1K	4P	0P7	ALG	TIB	BSX83	2N4352	0
3N175	NME	B56	30V	50MA	200C	.22WF	2MXV			200R	5P		RLS	OBS		3N169	0
3N176	NME	B56	25V	50MA	200C	.22WF	2.5MXV			300R	5P		RLS	OBS		2N4351	0
3N177	NME	B56	20V				3.5MXV			600R	7P		RLS	OBS		2N4351	0
3N178	PME	B53	75V	20MA	200C	0.1WF	5.5MXV	3MNMA		750R	3P5		RLS	GIB	BSW95	3N164	0
3N179	PME	B53	60V	20MA	200C	0.1WF	6MXV	3MNMA		1K	4P5		RLS	GIB	BSW95	3N164	0
3N180	PME	B53	40V	20MA	200C	0.1WF	6MXV	3MNMA		1K2	5P		RLS	GIB	BSW95	3N164	0
3N181	PME	B53	30V	100MA	200C	0.3WF	4MXV			40R	25P		RLS	GIB		3N168	0
3N182	PME	B53	30V	100MA	200C	0.3WF	5MXV			60R	25P		RLS	GIB		3N168	0
3N183	PME	B53	25V	100MA	200C	0.3WF	6MXV			75R	30P		RLS	GIB		3N168	0
3N184	PME	B53	30V	50MA	200C	0.3WF	3MXV	20MNMA		150R	9P		RLS	GIB		3N161	0
3N185	PME	B53	30V	50MA	200C	0.3WF	6MXV	15MNMA		175R	10P		RLS	GIB		3N161	0
3N186	PME	B53	25V	50MA	200C	0.3WF	3.5MXV	10MNMA		200R	11P		RLS	GIB		3N161	0
3N187	NMD	B66	20V	50MA	175C	.33WF	0.5/4V	5/30MA	7/18MO		8P5	0P03	TUG	RCB		3N225A	0
3N188	PME	B42	40V	50MA	150C	.52WF	2/5V	5/30MA	1.5/4MO	300R	4P5	1P5	DUA	INB		2N4067	0
3N189	PME	B42	40V	50MA	150C	.52WF	2/5V	5/30MA	1.5/4MO	300R	4P5	1P5	DUA	INB		2N4067	0
3N190	PME	B42	40V	50MA	150C	.52WF	2/5V	5/30MA	1.5/4MO	300R	4P5	1P	DUA	INB		2N4067	0
3N190/D	PME	B73	40V	50MA	150C	.52WF	2/5V	5/30MA	1.5/4MO	300R	4P5	1P	DUA	INB		2N4067	0
3N190/W	PME	B73	40V	50MA	150C	.52WF	2/5V	5/30MA	1.5/4MO	300R	4P5	1P	DUA	INB		2N4067	0
3N191	PME	B42	40V	50MA	150C	.52WF	2/5V	5/30MA	1.5/4MO	300R	4P5	1P	DUA	INB		2N4067	0

TYPE NO	CONS TRUC TION	CASE & LEAD	V DS MAX	I D MAX	T J MAX	P TOT MAX	VP OR VT	IDSS OR IDOM	G MO	R DS MAX	C ISS MAX	C RSS MAX	USE	SUPP LIER	EUR SUB	USA SUB	ISS
3N191/D	PME	B73	40V	50MA	150C	.52WF	2/5V	5/30MA	1.5/4MO	300R	4P5	1P	DUA	INB		2N4067	0
3N191/W	PME	B74	40V	50MA	150C	.52WF	2/5V	5/30MA	1.5/4MO	300R	4P5	1P	DUA	INB		2N4067	0
3N192	NMD	B54	20V	50MA	150C	.22WF	4MXV	30MXMA	8/24MO		6P		FVG	OBS	BF352	3N201	0
3N193	NMD	B54	20V	50MA	150C	.22WF	4MXV	20MXMA	6/22MO		7P		FVG	OBS	BF354	3N187	0
3N200	NMD	B66	20V	50MA	175C	.33WF	0.1/3V	0.5/12MA	10/20MO		8P5	OP03	TUG	RCU		3N206	0
3N201	NMD	B66	30V	50MA	200C	.36WF	0.5/5V	6/30MA	8/20MO		6P	OP03	FVG	SIU	BF352	3N187	0
3N202	NMD	B66	30V	50MA	200C	.36WF	0.5/5V	6/30MA	8/20MO		6P	OP03	FVG	SIU	BF352	3N187	0
3N203	NMD	B66	25V	50MA	200C	1.2WC	0.5/5V	3/15MA	7/15MO		6P	OP03	FVG	SIU	BF350	3N141	0
3N203A	NMD	B66	25V	50MA	200C	1.2WC	0.5/5V	3/15MA	7/15MO		6P	OP03	FVG	TIB	BF350	3N141	0
3N204	NMD	B66	25V	50MA	200C	.36WF	0.2/4V	6/30MA	10/22MO			OP03	TUG	TIB		3N225A	0
3N205	NMD	B66	25V	50MA	200C	.36WF	0.2/4V	6/30MA	10/22MO			OP03	TUG	TIB		3N225A	0
3N206	NMD	B66	25V	50MA	200C	.36WF	0.2/4V	3/15MA	7/17MO			OP03	TUG	TIB		3N225A	0
3N207	PME	B42	25V	100MA	200C	0.3WF	3/6V	1.5MNMA		400R	4P	2P5	DUA	TIB		3N208	0
3N208	PME	B42	25V	100MA	200C	0.3WF	6MXV	1.5MNMA		400R	4P	2P5	DUA	TIB		3N207	0
3N209	NMD	B66	25V	30MA	200C	0.3WF	4MXV	5/30MA	10/20MO		7P	OP03	TUG	MOB		3N204	0
3N210	NMD	B85	25V	30MA	150C	.35WF	4MXV	5/30MA	10/20MO		7P	OP03	TUG	MOB			0
3N211	NMD	B66	27V	50MA	175C	.36WF	0.5/5.5V	6/40MA	17/40MO			OP05	FVG	RCB			0
3N212	NMD	B66	27V	50MA	175C	.36WF	0.5/4V	6/40MA	17/40MO			OP05	FVG	RCB			0
3N213	NMD	B66	35V	50MA	175C	.36WF	0.5/5.5V	6/40MA	15/35MO			OP05	FVG	RCB			0
3N214	NMD	B15	20V	50MA	200C	.36WF		50MNMA		20R	6P	2P	RLS	TIB			0
3N215	NMD	B15	20V	50MA	200C	.36WF		50MNMA		35R	6P	2P	RLS	TIB		3N214	0
3N216	NMD	B15	20V	50MA	200C	.36WF		50MNMA		50R	6P	2P	RLS	TIB		3N215	0
3N217	NMD	B15	20V	50MA	200C	.36WF		50MNMA		70R	6P	2P	RLS	TIB		3N216	0
3N218	PME	B53	25V	700MA	175C	0.3WF	3.5MXV			20R	50P		RMS	GIU		3N167	0
3N223	NMD	B66	30V	50MA	200C	0.3WF	2MXV	12MXMA	17/40MO				FVG	MOU		3N211	0

TYPE NO	CONS TRUC TION	CASE & LEAD	V DS MAX	I D MAX	T J MAX	P TOT MAX	VP OR VT	IDSS OR IDOM	G MO	R DS MAX	C ISS MAX	C RSS MAX	USE	SUPP LIER	EUR SUB	USA SUB	ISS
3N225	NMD	B18	20V	20MA	200C	0.3WF	4MXV	1/20MA	6/15MO			0P03	TUG	TIB		3N204	0
3N225A	NMD	B18	20V	20MA	200C	0.3WF	4MXV	1/15MA	7.5/15MO			0P03	TUG	TIB		3N204	0
3SJ11	PME	B19	30V	10MA	150C	0.1WF	6.5MXV		0.3MNMO	1K	5P		RLS	NIP			0
3SJ11A	PME	B56	30V	50MA	150C	.22WF	3.5MXV		0.5MNMO		8P		ALG	NIP		3N158A	0
3SK14	NJD	B12	20V		150C	0.1WF	5MXV						ALG	OBS	BFW61	2N3819	0
3SK15	NMD	B57	25V	10MA	150C	0.1WF	2MXV	5MXMA	0.5MNMO		4P		ALG	OBS	BFX63	2N3631	0
3SK15A	NMD	B57	25V	10MA	150C	0.1WF	2MXV	5MXMA	0.5MNMO		4P		ALG	OBS	BFX63	2N3631	0
3SK16	NMD	B57	25V	10MA	150C	0.1WF	2MXV	5MXMA	0.5MNMO		4P		ALG	OBS	BFX63	2N3631	0
3SK17	NMD	B57	15V	10MA	150C	0.1WF	2MXV	5MXMA	0.5MNMO		4P		ALG	OBS	BFX63	2N3631	0
3SK18	NMD	B57	15V	10MA	150C	0.1WF	2MXV	5MXMA	0.5MNMO		4P		ALG	OBS	BFX63	2N3631	0
3SK19	NMD	B57	15V	10MA	150C	0.1WF	2MXV	5MXMA	0.5MNMO		4P		ALG	OBS	BFX63	2N3631	0
3SK20H	NMD	B12	20V	10MA	150C	0.2WF		5MXMA	0.6/4.5MO		5P		ALG	HIB	BFX63	2N3631	0
3SK21H	NMD	B12	20V	10MA	150C	0.2WF		5MXMA	2.5MNMO		5P		ALG	HIB	BFX63	2N3631	0
3SK22	NJD	B54	18V	25MA	150C	0.2WF	1.2/8V	3/24MA	7TPMO			0P6	FVG	TOB	BF256C	2N5247	0
3SK22/B	NJD	B54	18V	25MA	150C	0.2WF	1.2/8V	12/24MA	7TPMO			0P6	FVG	TOB	BF256C	2N5247	0
3SK22/G	NJD	B54	18V	25MA	150C	0.2WF	1.2/8V	6/14MA	7TPMO			0P6	FVG	TOB	BF256	2N5245	0
3SK22/Y	NJD	B54	18V	25MA	150C	0.2WF	1.2/8V	3/7MA	7TPMO			0P6	FVG	TOB	BF256A	2N4416	0
3SK23	NJD	B54	15V		150C	0.2WF	5.5MXV	24MXMA	6/12MO		4P5		RLS	TOB	BFW61	2N5397	0
3SK28	NJD	B64	18V	25MA	150C	0.2WF	1.2/5.5V	3.7/22MA	4.5/13MO		6P	0P6	FVG	TOB	BFW61	2N5397	0
3SK28/B	NJD	B64	18V	25MA	150C	0.2WF	1.2/5.5V	11/22MA	4.5/13MO		6P	0P6	FVG	TOB	BF256C	2N5247	0
3SK28/G	NJD	B64	18V	25MA	150C	0.2WF	1.2/5.5V	6.5/13MA	4.5/13MO		6P	0P6	FVG	TOB	BF256B	2N5245	0
3SK28/Y	NJD	B64	18V	25MA	150C	0.2WF	1.2/5.5V	4/7.5MA	4.5/13MO		6P	0P6	FVG	TOB	BF256A	2N5485	0
3SK29	NMD	B57	20V	10MA	100C	0.8WF	5MXV	1TPMA	0.3MNMO		4P		ALG	NIP			0
3SK32	NMD	B55	20V	15MA	150C	.17WF	2.5MXV		5/10MO		5P		FVG	MAT	BFS28	3N225A	0
3SK33	NMD	B54	25V	20MA	150C	.25WF	4MXV	15MXMA	4MNMO		3P		FVG	NIP		3N128	0

TYPE NO	CONS TRUC TION	CASE & LEAD	V DS MAX	I D MAX	T J MAX	P TOT MAX	VP OR VT	IDSS OR IDOM	G MO	R DS MAX	C ISS MAX	C RSS MAX	USE	SUPP LIER	EUR SUB	USA SUB	ISS
3SK35	NMX	B66	20V	30MA	150C	0.3WF	4MXV	24MXMA	10TPMO		5P5	0P04	FVG	TOB	BF350	3N140	0
3SK35/B	NMX	B66	20V	30MA	150C	0.3WF	4MXV	12/24MA	10TPMO		5P5	0P04	FVG	TOB	BF352	3N201	0
3SK35/G	NMX	B66	20V	30MA	150C	0.3WF	4MXV	6/14MA	10TPMO		5P5	0P04	FVG	TOB	BF354	3N203	0
3SK35/Y	NMX	B66	20V	30MA	150C	0.3WF	4MXV	3/7MA	10TPMO		5P5	0P04	FVG	TOB	BFS28	3N203	0
3SK37	NMD	B66	20V	30MA	150C	.23WF	3MXV	10TPMA	7MNMO		4P		FVG	SOY	BFS28	3N201	0
3SK38	NME	B54	20V	10MA	125C	0.2WF	3MXV		0.35MNMO	500R	2P5		RLS	OBS		3N170	0
3SK38A	NME	B25	20V	10MA	125C	0.2WF	3MXV		0.35MNMO	500R	2P5		ALC	TOS		3N170	0
3SK39	NMD	B66	20V	24MA	125C	.25WF	3MXV	24MXMA	7/18MO		4P5		FVG	MAT	BFS28	3N187	0
3SK44	NMD	B18	20V	45MA	150C	0.3WF	0.3/3.3V	3/40MA	13TPMO		10P	0P08	FVG	TOB	BF352	3N211	0
3SK44/B	NMD	B18	20V	45MA	150C	0.3WF	0.3/3.3V	3/7MA	13TPMO		10P	0P08	FVG	TOB	BF350	3N203	0
3SK44/R	NMD	B18	20V	45MA	150C	0.3WF	0.3/3.3V	12/24MA	13TPMO		10P	0P08	FVG	TOB	BF352	3N202	0
3SK44/W	NMD	B18	20V	45MA	150C	0.3WF	0.3/3.3V	6/14MA	13TPMO		10P	0P08	FVG	TOS	BF354	3N203	0
3SK49	NMD	B66	20V	30MA	150C	.35WF		30MXMA	17TPMO		5P5		FVG	MAT	BFR84	3N209	0
3SK55	NMD	B66	20V	30MA	150C	0.3WF	2.5MXV	3/30MA	12MNMO		5P		FVG	TOS	BF352	3N211	0
3UT40	NJD	B25	20V	20MA	150C	0.2WF	2TPV	5MXMA	0.5MNMO		6P		ALG	OBS		3N125	0
4G2	NJD	B15	30V	16MA	200C	0.3WF	6.5MXV	16MXMA	4MNMO		7P5		ALG	OBS	BF810	2N5459	0
5G2	NJD	B15	30V	16MA	200C	0.3WF	4MXV	8MXMA	2MNMO		7P5		ALG	OBS	BC264C	2N3684	0
6G2	NJD	B15	30V	4MA	200C	0.3WF	3MXV	4MXMA	1MNMO		7P5		ALG	OBS	BF808	2N5457	0
286/1BFY	NMD	B54	30V	20MA	125C	0.2WF	5MXV	10MXMA	7MNMO				ALG	OBS	BFR29	3N128	0
528BSY	NME	B54	35V	50MA	125C	.15WF	6MXV		6TPMO	100R			RLS	OBS		3N170	0
40460	NMD	B56	32V	15MA	85C	.15WF				800R			RLS	OBS		3N138	0
40461	NMD	B56	25V	15MA	125C	.15WF		14MXMA	3.5TPMO		5P		ALG	OBS		2N3631	0
40467	NMD	B56	20V	50MA	125C	0.1WF	8MXV	10/50MA	4MNMO			0P28	FVG	OBS		3N128	0
40467A	NMD	B54	20V	50MA	175C	.33WF	8MXV	10/50MA	4MNMO			0P28	FVG	RCB		3N128	0
40468	NMD	B54	20V	30MA	125C	0.1WF	8MXV	5/30MA	7.5TPMO		5P5	0P28	FVG	OBS		3N142	0

TYPE NO	CONS TRUC TION	CASE & LEAD	V DS MAX	I D MAX	T J MAX	P TOT MAX	VP OR VT	IDSS OR IDOM	G MO	R DS MAX	C ISS MAX	C RSS MAX	USE	SUPP LIER	EUR SUB	USA SUB	ISS
40468A	NMD	B54	20V	30MA	175C	.38WF	8MXV	5/30MA	7.5TPMO		5P5	OP28	FVG	RCB		3N142	0
40559	NMD	B54	20V	50MA	175C	0.4WF	1/8V	5/30MA	7.5TPMO		5P5	OP28	FVG	OBS		3N128	0
40559A	NMD	B54	20V	30MA			1/8V	5/30MA	7.5TPMO		5P5	OP28	FVG	RCB		3N128	0
40600	NMD	B66	20V	50MA	175C	.33WF	3MXV	18TPMA	10TPMO		5P5	OP03	FVG	RCB		3N140	0
40601	NMD	B66	20V	50MA	175C	.33WF	3MXV	18TPMA	10TPMO		5P5	OP03	FVG	RCB		3N141	0
40602	NMD	B66	20V	50MA	175C	.33WF	3MXV	18TPMA	10TPMO		5P5	OP03	FVG	RCB		3N159	0
40603	NMD	B66	20V	50MA	175C	.33WF	3MXV	18TPMA	10TPMO		5P5	OP03	FVG	RCB		3N159	0
40604	NMD	B66	20V	50MA	175C	.33WF	3MXV	18TPMA	10TPMO		5P5	OP03	FVG	RCB		3N159	0
40673	NMD	B66	20V	50MA	175C	.33WF	4MXV	5/35MA	12TPMO		6P	OP03	TUG	RCB		3N209	0
40819	NMD	B66	25V	50MA	175C	.33WF	4MXV	5/35MA	12TPMO		8P5	OP03	TUG	RCB		3N209	0
40820	NMD	B66	20V	50MA	175C	.33WF	3MXV	0.5/15MA	12TPMO		8P5	OP03	TUG	RCB		3N206	0
40821	NMD	B66	20V	50MA	175C	.33WF	3MXV	0.5/20MA	12TPMO		9P	OP04	TUG	RCB		3N203	0
40822	NMD	B66	18V	50MA	175C	.33WF	4MXV	5/30MA	12TPMO		9P5	OP03	TUG	RCB		3N206	0
40823	NMD	B66	18V	50MA	175C	.33WF	4MXV	5/35MA	12TPMO		10P	OP05	TUG	RCB		3N203	0
40841	NMD	B66	18V	50MA	175C	0.3WF	2MXV	10TPMA	12MNMO		11P		FVG	OBS	BFS28	3N206	0
41004	PME	B80	30V	35MA	150C	.35WF	4MXV		0.8MNMO		8P		RLS	RCB		3N164	0
A190	NJD	B15	30V	20MA	175C	0.3WF	4MXV	8MXMA	2/8MO		5P		TUG	OBS	BF256A	2N5246	0
A191	NJD	B15	20V	10MA	175J	.15WF	4MXV	10MXMA	2/8MO		5P		TUG	OBS	BF256B	2N4416	0
A192	NJD	B15	30V	20MA	175C	0.3WF	8MXV	20MXMA	3.5/6.5MO		6P		TUG	OBS	BF256	2N5247	0
A193	NJD	B15	30V	20MA	175C	0.3WF	6MXV	10MXMA	1.5/6.5MO		6P		TUG	OBS	BF256B	2N5246	0
A194	NJD	B58	25V	15MA	150C	.15WF	4MXV	8MXMA	2/8MO		5P		TUG	OBS	BF256A	2N5246	0
A195	NJD	B58	25V	15MA	150C	.15WF	4MXV	6MXMA	1/6MO		5P		TUG	OBS	BF256A	2N5246	0
A196	NJD	B58	25V	15MA	150C	.15WF	4MXV	15MXMA	4/10MO		5P		TUG	OBS	BF256	2N5245	0
A197	NJD	B58	30V	150MA	150C	0.2WF	10MXV	150MXMA		30R	16P		RLS	OBS	BSV78	2N4391	0
A198	NJD	B58	30V	150MA	150C	0.2WF	5MXV	75MXMA		60R	16P		RLS	OBS	BSV79	2N4861	0

TYPE NO	CONS TRUC TION	CASE & LEAD	V DS MAX	I D MAX	T J MAX	P TOT MAX	VP OR VT	IDSS OR IDOM	G MO	R DS MAX	C ISS MAX	C RSS MAX	USE	SUPP LIER	EUR SUB	USA SUB	ISS
A199	NJD	B58	30V	150MA	150C	0.2WF	3MXV	30MXMA		100R	16P		RLS	OBS	BSV80	2N4093	0
A390	NME	OBS	25V	30MA	175C	0.3WF	5MXV			300R	5P5		RLS	OBS		3N169	0
A392	NME	B54	30V	50MA	125C	.15WF	5MXV			100R	5P		RLS	OBS		3N170	0
A498	NMD	B66	20V	20MA	125C	0.2WF	5MXV		0.8/1.2MO		4P5		ALG	OBS	BFX63	2N3631	0
A5T3821	NJD	B8	50V	25MA	125C	0.3WF	4MXV	.5/2.5MA	1.5/4.5MO		6P	3P	ALG	TIB	BF347	2N3821	0
A5T3822	NJD	B8	50V	25MA	125C	0.3WF	6MXV	2/10MA	3/6.5MO		6P	3P	ALG	TIB	BF244	2N3822	0
A5T3823	NJD	B8	30V	25MA	125C	0.3WF	8MXV	4/20MA	3.5/6.5MO		6P	2P	FVG	TIB	BFW61	2N3823	0
A5T3824	NJD	B8	50V	25MA	125C	0.3WF		12/24MA		250R	6P	3P	RLS	TIB		2N3824	0
A5T5460	PJD	B8	40V	10MA	150C	0.3WF	6MXV	5MXMA	1/4MO	2K	7P		ALG	TIB	BF320A	2N5460	0
A5T5461	PJD	B8	40V	10MA	150C	0.3WF	7.5MXV	9MXMA	1.5/5MO	2K	7P		ALG	TIB	BF320B	2N5461	0
A5T5462	PJD	B8	40V	10MA	150C	0.3WF	9MXV	16MXMA	2/6MO	2K	7P		ALG	TIB	BF320C	2N5462	0
A5T6449	NJD	B36	300V	10MA	150C	.62WF	2/15V	2/10MA	0.5/3MO		10P	5P	ALH	TIB		2N5543	0
A5T6450	NJD	B36	200V	10MA	150C	0.6W	2/15V	2/10MA	0.5/3MO		10P	5P	ALH	TIB		2N5543	0
A610L	NJD	B15	30V	10MA	200C	.15WF	4MXV	8MXMA	2/8MO		5P		MPP	OBS	BFS21A	MMF1	0
A610S	NJD	B15	30V	10MA	200C	.15WF	4MXV	8MXMA	2/8MO		5P		MPP	OBS	BFS21A	MMF1	0
A611L	NJD	B15	30V	10MA	200C	.15WF	4MXV	8MXMA	2/8MO		5P		MPP	OBS	BFS21A	MMF1	0
A611S	NJD	B15	30V	10MA	200C	.15WF	4MXV	8MXMA	2/8MO		5P		MPP	OBS	BFS21A	MMF1	0
AD830	NJD	B41	40V	0.5MA	150C	0.5WF	0.6/4.5V	0.03MNMA	.07/.25MO		3P	1P5	DUA	AND		2N5906	0
AD831	NJD	B41	40V	0.5MA	150C	0.5WF	0.6/4.5V	0.03MNMA	0.7/.25MO		3P	1P5	DUA	AND		2N5907	0
AD832	NJD	B41	40V	0.5MA	150C	0.5WF	0.6/4.5V	0.03MNMA	.07/.25MO		3P	1P5	DUA	AND		2N5908	0
AD833	NJD	B41	40V	0.5MA	150C	0.5WF	0.6/4.5V	0:03MNMA	.07/.25MO		3P	1P5	DUA	AND		2N5909	0
AD833A	NJD	B41	40V	0.5MA	150C	0.5WF	0.6/4.5V	0.03MNMA	.07/.25MO		3P	1P5	DUA	AND		2N5909	0
AD835	NJD	B51	30V	5MA	200C	0.5WF	3.5MXV	0.5MNMA	1.5TPMO				DUA	AND		2N5906	0
AD836	NJD	B51	30V	5MA	200C	0.5WF	3.5MXV	0.5MNMA	1.5TPMO				DUA	AND		2N5906	0
AD837	NJD	B51	30V	5MA	200C	0.5WF	3.5MXV	0.5MNMA	1.5TPMO				DUA	AND		2N5906	0

TYPE NO	CONS TRUC TION	CASE & LEAD	V DS MAX	I D MAX	T J MAX	P TOT MAX	VP OR VT	IDSS OR IDOM	G MO	R DS MAX	C ISS MAX	C RSS MAX	USE	SUPP LIER	EUR SUB	USA SUB	ISS
AD838	NJD	B51	30V	5MA	200C	0.5WF	3.5MXV	0.5MNMA	1.5TPMO				DUA	AND		2N5906	0
AD839	NJD	B51	30V	5MA	200C	0.5WF	3.5MXV	0.5MNMA	1.5TPMO				DUA	AND		2N5906	0
AD840	NJD	B51	40V	5MA	200C	0.5WF	4.5MXV	0.5MNMA	0.8TPMO				DUA	AND		2N5520	0
AD841	NJD	B51	40V	5MA	200C	0.5WF	4.5MXV	0.5MNMA	0.8TPMO				DUA	AND		2N5521	0
AD842	NJD	B51	40V	5MA	200C	0.5WF	4.5MXV	0.5MNMA	0.8TPMO				DUA	AND		2N5523	0
AD3954	NJD	B51	40V	5MA	125C	0.5WF	1/4.5V	0.5/5MA	1/3MO		4P	1P2	DUA	AND	BFQ10	2N3954	0
AD3954A	NJD	B51	40V	5MA	125C	0.5WF	1/4.5V	0.5/5MA	1/3MO		4P	1P2	DUA	AND	BFQ10	2N3954A	0
AD3955	NJD	B51	40V	5MA	125C	0.5WF	1/4.5V	0.5/5MA	1/3MO		4P	1P2	DUA	AND	BFQ10	2N3955	0
AD3955A	NJD	B51	40V	5MA	125C	0.5WF	1/4.5V	5MA	1/3MO		4P	1P2	DUA	AND	BFQ10	2N3955A	0
AD3956	NJD	B51	40V	5MA	125C	0.5WF	1/4.5V	0.5/5MA	1/3MO		4P	1P2	DUA	AND	BFQ10	2N3956	0
AD3957	NJD	B51	40V	5MA	125C	0.5WF	1/4.5V	0.5/5MA	1/3MO		4P	1P2	DUA	AND	BFQ10	2N3957	0
AD3958	NJD	B51	50V	5MA	125C	0.5WF	1/4.5V	0.5/5MA	1/3MO		4P	1P2	DUA	AND	BFQ10	2N3958	0
AD5905	NJD	B43	40V	0.5MA	150C	0.5WF	0.6/4.5V	.03/.5MA	.07/.25MO		3P	1P5	DUA	AND		2N5905	0
AD5906	NJD	B43	40V	0.5MA	150C	0.5WF	0.6/4.5V	.03/.5MA	.07/.25MO		3P	1P5	DUA	AND		2N5906	0
AD5907	NJD	B43	40V	0.5MA	150C	0.5WF	0.6/4.5V	.03/.5MA	.07/.25MO		3P	1P5	DUA	AND		2N5907	0
AD5908	NJD	B43	40V	0.5MA	150C	0.5WF	0.6/4.5V	.03/.5MA	.07/.25MO		3P	1P5	DUA	AND		2N5908	0
AD5909	NJD	B43	40V	0.5MA	150C	0.5WF	0.6/4.5V	.03/.5MA	.07/.25MO		3P	1P5	DUA	AND		2N5909	0
AE2211N	NJD	B41	20V	6MA	200C	0.6WF	0.5TPV	0.15MXMA	0.3TPMO				DUA	OBS		2N5902	0
AE2212N	NJD	B41	20V	6MA	200C	0.6WF	0.7TPV	0.1/.3MA	0.5TPMO				DUA	OBS		2N5902	0
AE2213N	NJD	B41	20V	6MA	200V	0.6WF	1TPV	0.2/.6MA	0.7TPMO				DUA	OBS		2N5902	0
AE2214N	NJD	B41	20V	6MA	200C	0.6WF	1.5TPV	.5/1.5MA	1.2TPMO				DUA	OBS		2N5902	0
AE2215N	NJD	B41	20V	6MA	200C	0.6WF	2.5TPV	1/3MA	1.6TPMO				DUA	OBS		2N4082	0
AE2216N	NJD	B41	20V	6MA	200C	0.6WF	3.5TPV	2/6MA	2.1TPMO				DUA	OBS		2N3954	0
BC264	NJD	B36	30V	60MA	150C	0.3WF	0.5/8V	2/12MA	2.5MNMO		6P	2P	ALN	MUB	BFW11	2N3684	0
BC264A	NJD	B36	30V	60MA	150C	0.3WF	0.5/8V	2/4.5MA	2.5MNMO		6P	2P	ALN	MUB	BFW56	2N3684	0

TYPE NO	CONS TRUC TION	CASE & LEAD	V DS MAX	I D MAX	T J MAX	P TOT MAX	VP OR VT	IDSS OR IDOM	G MO	R DS MAX	C ISS MAX	C RSS MAX	USE	SUPP LIER	EUR SUB	USA SUB	ISS
BC264B	NJD	B36	30V	60MA	150C	0.3WF	0.5/8V	3.5/7MA	2.5MNMO		6P	2P	ALN	MUB	BFW11	2N3684A	0
BC264C	NJD	B36	30V	60MA	150C	0.3WF	0.5/8V	5/8MA	2.5MNMO		6P	2P	ALN	MUB	BFW11	2N6451	0
BC264D	NJD	B36	30V	60MA	150C	0.3WF	0.5/8V	7/12MA	2.5MNMO		6P	2P	ALN	MUB	BFW11	2N3684A	0
BC264L	NJD	B1	30V	60MA	150C	0.3WF	0.5/8V	2/12MA	2.5MNMO		6P	2P	ALN	TIB	BFW11	2N3684	0
BC264LA	NJD	B1	30V	60MA	150C	0.3WF	0.5/8V	2/4.5MA	2.5MNMO		6P	2P	ALN	TIB	BFW54	2N3684	0
BC264LB	NJD	B1	30V	60MA	150C	0.3WF	0.5/8V	3.5/7MA	2.5MNMO		6P	2P	ALN	TIB	BFW11	2N3684	0
BC264LC	NJD	B1	30V	60MA	150C	0.3WF	0.5/8V	5/8MA	2.5MNMO		6P	2P	ALN	TIB	BFW11	2N6451	0
BC264LD	NJD	B1	30V	60MA	150C	0.3WF	0.5/8V	7/12MA	2.5MNMO		6P	2P	ALN	TIB	BFW11	2N3684	0
BF244	NJD	B1	30V	MA	150C	0.3WF	0.5/8V	2/25MA	3/6.5MO		4P	1P1	ALG	MUB	BFW61	2N3819	0
BF244A	NJD	B1	30V	10MA	125C	0.3WF	0.510.8V	2/6.5MA	3/6.5MO		8P	2P2	ALG	MUB	BFW61	2N5458	0
BF244B	NJD	B1	30V	10MA	125C	0.3WF	0.5/0.8V	6/15MA	3/6.5MO		8P	2P2	ALG	MUB	BC264D	2N6451	0
BF244C	NJD	B1	30V	10MA	125C	0.3WF	0.5/0.8V	12/25MA	3/6.5MO		8P	2P2	ALG	MUB	BF347	2N6451	0
BF245	NJD	B36	30V	35MA	150C	0.3WF	0.5/8V	2/25MA	3/6.5MO		8P	2P	ALG	MUB	BF811	2N6451	0
BF245A	NJD	B36	30V	35MA	150C	0.3WF	0.5/8V	2/6.5MA	3/6.5MO		8P	2P	ALG	MUB	BFW56	2N3822	0
BF245B	NJD	B36	30V	35MA	150C	0.3WF	0.5/8V	6/15MA	3/6.5MO		8P	2P	ALG	MUB	BF811	2N5459	0
BF245C	NJD	B36	30V	35MA	150C	0.3WF	0.5/8V	12/25MA	3/6.5MO		8P	2P	ALG	MUB	BF244	2N5397	0
BF246	NJD	B1	25V	260MA	150C	0.3WF	.6/14.5V	30/250MA	8/50MO		30P	7P	FVG	TIB	BSV78	2N4859	0
BF246A	NJD	B1	25V	260MA	150C	0.3WF	.6/14.5V	30/80MA	8/50MO		30P	7P	FVG	TIB	BSV78	2N4392	0
BF246B	NJD	B1	25V	260MA	150C	0.3WF	.6/14.5V	60/140MA	8/50MO		30P	7P	FVG	TIB	BSV78	2N4856	0
BF246C	NJD	B1	25V	260MA	150C	0.3WF	0.6/14V	.11/.25A	8/50MO		30P	7P	FVG	TIB	BSV78	2N4856	0
BF247	NJD	B36	30V	260MA	150C	0.3WF	0.6/15V	30/250MA	8/50MO		30P	7P	FVG	TIB	BF348	2N6453	0
BF247A	NJD	B36	25V	260MA	150C	0.3WF	0.6/14V	30/80MA	8/50MO		30P	7P	FVG	TIB	BSV78	2N4392	0
BF247B	NJD	B36	25V	260MA	150C	0.2WF	0.6/14V	60/140MA	8/50MO		30P	7P	FVG	TIB	BSV78	2N4856	0
BF247C	NJD	B36	25V	260MA	150C	0.2WF	0.6/14V	.11/.25A	8/50MO		30P	7P	FVG	TIB	BSV78	2N4856	0
BF256	NJD	B36	30V	30MA	150C	0.3WF	0.5/7.5V	3/18MA	4.5MNMO		4P5	1P2	TUG	MUB	BF256L	2N5245	0

TYPE NO	CONS TRUC TION	CASE & LEAD	V DS MAX	I D MAX	T J MAX	P TOT MAX	VP OR VT	IDSS OR IDOM	G MO	R DS MAX	C ISS MAX	C RSS MAX	USE	SUPP LIER	EUR SUB	USA SUB	ISS
BF256A	NJD	B36	30V	10MA	150C	0.3WF	0.5/7.5V	3/7MA	4.5MNMO		4P5	1P2	TUG	MUB	BF256LA	2N5246	0
BF256B	NJD	B36	30V	30MA	150C	0.3WF	0.5/7.5V	6/13MA	4.5MNMO		4P5	1P2	TUG	MUB	BF256LB	2N5245	0
BF256C	NJD	B36	30V	30MA	150C	0.3WF	0.5/7.5V	11/18MA	4.5MNMO		4P5	1P2	TUG	MUB	BF256LC	2N4416	0
BF256L	NJD	B1	30V	30MA	150C	0.3WF	0.5/7.5V	3/18MA	4.5MNMO		4P5	1P2	TUG	MUB	BF256	2N5245	0
BF256LA	NJD	B1	30V	10MA	150C	0.3WF	0.5/7.5V	3/7MA	4.5MNMO		4P5	1P2	TUG	TIB	BF256A	2N5246	0
BF256LB	NJD	B1	30V	30MA	150C	0.3WF	0.5/7.5V	6/13MA	4.5MNMO		4P5	1P2	TUG	TIB	BF256B	2N5245	0
BF256LC	NJD	B1	30V	30MA	150C	0.3WF	0.5/7.5V	11/18MA	4.5MNMO		4P5	1P2	TUG	TIB	BF256C	2N4416	0
BF320	PJD	B1	15V	25MA	125C	0.2WF	8MXV	0.3/15MA	0.8/5MO		32P	16P	ALG	TIB	BFX83	2N3380	0
BF320A	PJD	B1	15V	25MA	125C	0.2WF	8MXV	.3/4.5MA	0.8/5MO		32P	16P	ALG	TIB	BFX83	2N5270	0
BF320B	PJD	B1	15V	25MA	125C	0.2WF	8MXV	3/8.5MA	0.8/5MO		32P	16P	ALG	TIB	BFX83	2N3380	0
BF320C	PJD	B1	15V	25MA	125C	0.2WF	8MXV	7.5/15MA	0.8/5MO		32P	16P	ALG	TIB	BFX83	2N3380	0
BF346	NJD	B36	15V	90MA	135C	0.3WF	0.5/5.5V	2/80MA	16/60MO			5P6	ALG	TIW	BF818	2N6453	0
BF346A	NJD	B36	15V	90MA	135C	0.3WF	0.5/5.5V	2/18MA	16/60MO			5P6	ALG	TIW	BF811	2N6451	0
BF346B	NJD	B36	15V	90MA	135C	0.3WF	0.5/5.5V	12/35MA	16/60MO			5P6	ALG	TIW	BF244	2N6453	0
BF346C	NJD	B36	15V	90MA	135C	0.3WF	0.5/5.5V	28/80MA	16/60MO			5P6	ALG	TIW	BF818	2N6453	0
BF347	NJD	B1	30V	12MA	150C	0.4WF	1MXV	.5/2.5MA	3/7.5MO	350R			RLS	TIB	BF256A	2N3687	0
BF348	NJD	B2	40V	25MA	150C	0.2WF	6MXV	10/60MA	10/15MO			1P5	FVG	TIW	BFW61	2N5397	0
BF350	NMD	B18	15V	50MA	175C	0.4WF	5MXV	3/30MA	10TPMO		20P	.04P	FVG	TIW	BFS28	3N140	0
BF351	NMD	B18	24V	50MA	175C	0.4WF	1/5V	5/30MA	14TPMO		12P	.04P	FVG	TIW	BFS28	3N140	0
BF352	NMD	B18	24V	50MA	175C	0.4WF	2MXV	5/30MA	14TPMO		12P	.04P	FVG	TIW	BF353	3N187	0
BF353	NMD	B18	24V	50MA	175C	0.4WF	3MXV	5/30MA	12TPMO		12P	.02P	FVG	TIW	BF905	3N140	0
BF354	NMD	B18	24V	50MA	175C	0.4WF	3MNV	7/15MA	1.5/3.5MO		3P5	0P02	FVG	TIW	BFS28	3N187	0
BF800	NJD	B15	25V	10MA	200C	0.4WF	6MXV	.3/1.2MA	0.2/0.7MO		2P6	0P8	ALN	TIB	BF801	2N4867	0
BF801	NJD	B15	25V	10MA	200C	0.4WF	6MXV	.3/1.2MA	0.2/0.7MO		2P6	0P8	ALN	TIB	BF801	2N4867	0
BF802	NJD	B15	25V	10MA	200C	0.4WF	6MXV	.3/1.2MA	0.2/0.7MO		2P6	0P8	ALN	TIB	BF800	2N3687A	0

TYPE NO	CONS TRUC TION	CASE & LEAD	V DS MAX	I D MAX	T J MAX	P TOT MAX	VP OR VT	.DSS OR IDOM	G MO	R DS MAX	C ISS MAX	C RSS MAX	USE	SUPP LIER	EUR SUB	USA SUB	ISS
BF803	NJD	B15	30V	1.2MA	200C	0.4WF	3MXV	0.8MXMA	0.25/.5MO		2P6	0P8	ALN	TIB		2N3687A	0
BF804	NJD	B15	30V	1.2MA	200C	0.4WF	5MXV	.3/1.2MA	0.25/.5MO		2P6	0P8	ALN	TIB	BF800	2N3687A	0
BF805	NJD	B15	30V	10MA	200C	0.3WF	6MXV	3/13MA	0.3/0.5MO		3P5	1P5	ALN	TIB	BF806	2N4869	0
BF806	NJD	B15	30V	10MA	200C	0.3WF	6MXV	3/13MA	3/5.5MO		3P5	1P5	ALN	TIB	BF805	2N4869	0
BF808	NJD	B15	20V	10MA	200C	.12WF	5MXV	1/6MA	0.8/2.5MO		2P5	0P4	ALN	TIB	BFW56	2N4868	0
BF810	NJD	B15	30V	10MA	200C	0.3WF	6MXV	5/20MA	5/9MO		6P	1P8	ALN	TIB	BF811	2N6451	0
BF811	NJD	B15	30V	10MA	200C	0.3WF	6MXV	5/20MA	5/9MO		6P	1P8	ALN	TIB	BF810	2N6451	0
BF815	NJD	B15	30V	10MA	200C	0.3WF	6MXV	15/40MA	10/20MO		10P	2P8	ALN	TIB	BF817	2N6451	0
BF816	NJD	B15	30V	10MA	200C	0.3WF	6MXV	15/40MA	10/20MO		10P	2P8	ALN	TIB	BF815	2N6451	0
BF817	NJD	B15	25V	10MA	200C	0.3WF	5MXV	10/40MA	15/25MO		15P	4P	ALN	TIB	BF815	2N6453	0
BF818	NJD	B15	25V	10MA	200C	0.3WF	5MXV	10/40MA	15/25MO		15P	4P	ALN	TIB	BF817	2N6451	0
BF960	NMD	B78	20V	30MA	125C	.15WF			9TPMO		20P	1P	TUG	SID	BF905		0
BFQ10	NJD	B20	30V	30MA	200C	0.2WF	0.5/3.5V	0.5/10MA	1MNMO		8P	1P	DUA	MUB		2N5902	0
BFQ11	NJD	B20	30V	30MA	200C	0.2WF	0.5/3.5V	0.5/10MA	1MNMO		8P	1P	DUA	MUB	BFQ10	2N5902	0
BFQ12	NJD	B20	30V	30MA	200C	0.2WF	0.5/3.5V	0.5/10MA	1MNMO		8P	1P	DUA	MUB	BFQ11	2N5903	0
BFQ13	NJD	B20	30V	30MA	200C	0.2WF	0.5/3.5V	0.5/10MA	1MNMO		8P	1P	DUA	MUB	BFQ12	2N5904	0
BFQ14	NJD	B20	30V	30MA	200C	0.2WF	0.5/3.5V	0.5/10MA	1MNMO		8P	1P	DUA	MUB	BFQ13	2N5904	0
BFQ15	NJD	B20	30V	30MA	200C	0.2WF	0.5/3.5V	0.5/10MA	1MNMO		8P	1P	DUA	MUB	BFQ14	2N5905	0
BFQ16	NJD	B20	30V	30MA	200C	0.2WF	0.5/3.5V	0.5/10MA	1MNMO		8P	1P	DUA	MUB	BFQ15	2N5905	0
BFR29	NMD	B14	30V	50MA	125C	0.2WF	4MXV	10/40MA	6MNMO		5P	0P7	ALN	MUB	BFX63	2N3631	0
BFR30	NJD	B59	25V	10MA	125C	0.1WF	5MXV	4/10MA	1/4MO		4P	1P5	ALG	MUB			0
BFR31	NJD	B59	25V	10MA	125C	0.1WF	2.5MXV	1/5MA	1.5/4.5MO		4P	1P5	ALG	MUB			0
BFR45	NJD	B15	30V	25MA	200C	0.3WF	0.5/8V	2/25MA	3/6.5MO		80P	2P2	ALG	TIW	BFW61	2N3822	0
BFR46	PME	B6	15V	10MA	200C	0.3WF	2TPV		1.7TPMO				ALN	FEB	BSW95	3N157	0
BFR47	PME	B6	15V	10MA	200C	0.3WF	2TPV		1.7TPMO				ALN	FEB	BSW95	3N157	0

TYPE NO	CONS TRUC TION	CASE & LEAD	V DS MAX	I D MAX	T J MAX	P TOT MAX	VP OR VT	IDSS OR IDOM	G MO	R DS MAX	C ISS MAX	C RSS MAX	USE	SUPP LIER	EUR SUB	USA SUB	ISS
BFR84	NMD	B18	20V	20MA	135C	0.2WF	4MXV		8MNMO			0P05	FVG	MUB	BF351	3N201	0
BFS21	NJD	B15	30V	20MA	200C	0.3WF	6MXV	1MNMA	1MNMO		5P	0P75	MPP	MUB		2N5199	0
BFS21A	NJD	B15	20V	20MA	200C	0.3WF	6MXV	1MNMA	1MNMO		5P	0P75	MPP	MUB	BFS21	2N5199	0
BFS21AX2	NJD	B15	20V	20MA	200C	0.3WF	6MXV	1MNMA	1MNMO		5P	0P75	MPP	MUB	BFS21X2	2N5199	0
BFS21X2	NJD	B15	30V	20MA	200C	0.3WF	6MXV	1MNMA	1MNMO		5P	0P75	MPP	MUB		2N5199	0
BFS28	NMD	B18	20V	20MA	135C	0.2WF	4MXV		8MNMO			0P05	FVG	MUB	BF351	3N201	0
BFS28R	NMD	B66	20V	20MA	135C	0.2WF	5MXV		8MNMO				FVG	OBS	BFR84	3N201	0
BFS67	NJD	B60	50V	50MA	150C	0.1WF	6MXV	0.5/10MA	1.5/6.5MO		6P	3P	ALG	TIB	BFS68		0
BFS67P	NJD	BB6	150V	10MA	150C	0.1WF	6MXV	0.5/10MA	1.5/6.5MO		6P	3P	ALG	TIB	BFS68P		0
BFS68	NJD	B60	30V	10MA	150C	0.1WF	8MXV	4/25MA	3.5/6.5MO		6P	3P	ALG	TIB	BFS67		0
BFS68P	NJD	B61	30V	10MA	150C	0.1WF	8MXV	4/25MA	3.5/6.5MO		6P	3P	ALG	TIB	BFS67P		0
BFS70	NJD	B15	30V	10MA	175C	0.3WF	4MXV	1/2.5MA	1.5/4.5MO		6P	3P	ALG	TIW	BFW12	2N3821	0
BFS71	NJD	B15	50V	10MA	175C	0.3WF	6MXV	2/10MA	3/6.5MO		6P	3P	ALG	TIW	BFW61	2N3821	0
BFS72	NJD	B15	30V	10MA	175C	0.3WF	8MXV	4/20MA	3.2/6.5MO		6P	2P	ALG	TIW	BFW61	2N3822	0
BFS73	NJD	B12	50V	10MA	175C	0.3WF					6P	3P	ALG	TIW	BFW61	2N3821	0
BFS74	NJD	B6	40V	50MA	175C	0.4WF	4/10V	50MNMA			18P	8P	RLS	TIW	BSV78	2N4091	0
BFS75	NJD	B6	40V	50MA	175C	0.4WF	2/6V	20/100MA			18P	8P	RLS	TIW	BSV79	2N4092	0
BFS76	NJD	B6	40V	50MA	175C	0.4WF	0.8/4V	8/80MA			18P	8P	RLS	TIW	BSV80	2N4093	0
BFS77	NJD	B6	30V	50MA	175C	0.4WF	4/10V	50MNMA			18P	8P	RLS	TIW	BSV78	2N4091	0
BFS78	NJD	B6	30V	50MA	175C	0.4WF	2/6V	20/100MA			18P	8P	RLS	TIW	BSV79	2N4092	0
BFS79	NJD	B6	30V	50MA	175C	0.4WF	0.8/4V	8/80MA			18P	8P	RLS	TIW	BSV80	2N4093	0
BFS80	NJD	B15	30V	10MA	175C	0.3WF	2.5/6V	5/15MA	4.5/7.5MO		4P	2P	ALG	TIW	BF810	2N3822	0
BFT10	NJD	B36	40V	25MA	150C	0.3WF		10MNMA	6/20MO		8P	3P	FVG	TIW	BF348	2N5397	0
BFT11	PJD	B36	25V	10MA	150C	0.3WF		10MNMA	6/15MO		18P	6P	FVG	TIW	BFT11	2N3384	0
BFW10	NJD	B15	30V	20MA	200C	0.3WF	8MXV	8/20MA	3.5/6.5MO		5P	0P8	FVG	MUB	BF348	2N5105	0

TYPE NO	CONS TRUC TION	CASE & LEAD	V DS MAX	I D MAX	T J MAX	P TOT MAX	VP OR VT	IDSS OR IDOM	G MO	R DS MAX	C ISS MAX	C RSS MAX	USE	SUPP LIER	EUR SUB	USA SUB	ISS
BFW11	NJD	B15	30V	20MA	200C	0.3WF	6MXV	4/10MA	3/6.5MO		5P	OP08	FVG	MUB	BF256	2N5245	0
BFW12	NJD	B36	30V	10MA	200C	.15WF	2.5MXV	1/5MA	2MNMO		5P	1P	FVG	MUB	BFW61	2N5103	0
BFW13	NJD	B36	30V	10MA	200C	.15WF	1.2MXV	.2/1.5MA	1MNMO		5P	1P	FVG	MUB	BFW13	2N4867	0
BFW54	NJD	B16	50V	10MA	200C	0.3WF	6MXV	2/10MA	3/6.5MO		6P	3P	ALN	TIB	BFW61	2N3822	0
BFW55	NJD	B16	50V	10MA	200C	0.3WF	6MXV	2/10MA	3/6.5MO		6P	3P	ALN	TIB	BFW61	2N3822	0
BFW56	NJD	B16	50V	10MA	200C	0.1WF	6MXV	2/10MA	3/6.5MO		6P	3P	ALG	TIB	BFW61	2N3822	0
BFW61	NJD	B15	25V	20MA	200C	0.3WF	8MXV	2/20MA	2/6.5MO		6P	2P	ALG	MUB	BFW61	2N3822	0
BFW96	NMD	B15	30V	50MA	200C	0.2WF	4.5MXV	30MXMA	1.3MNMO		6P	1P	ALG	VAL	BFX63	2N3631	0
BFX63	NMD	B24	30V	50MA	175C	0.2WF	4.5MNV	30MXMA	1.3MNMO		5P	OP8	FVG	MUB	BFW96	2N3631	0
BFX78	NJD	B52	15V	30MA	175C	.37WF		25MXMA	6/9MO		3P		FVG	SGS	BF810	2N5397	0
BFX82	PJD	B6	25V	30MA	175C	0.3WF	5MXV	12MXMA	2/6MO	0.3K	20P		RLS	OBS	BF320	2N2609	0
BFX83	PJD	B6	25V	30MA	175C	0.3WF	9MXV	30MXMA	4/8MO	0.2K	20P		RLS	OBS	BF320C	2N2609	0
BSV20	PME	B53	30V	20MA	175C	0.4WF	6MXV	16MXMA	4.6MNMO	250R	12P		RLS	OBS	BSX83	3N161	0
BSV20A	PME	B53	30V	20MA	175C	0.4WF	6MXV	16MXMA	4.6MNMO	250R	12P		RLS	OBS	BSX83	3N161	0
BSV22	NJD	B15	30V	50MA	200C	0.2WF		30MXMA			5P	OP8	ALG	TIB	BFW61	2N5397	0
BSV34	PME	B38	30V	200MA	175C	0.6WF	6MXV	8MXMA	2MNMO	500R	6P		DUA	OBS		3N207	0
BSV34A	PME	B38	30V	200MA	175C	0.6WF	6MXV	8MXMA	2MNMO	500R	6P		DUA	OBS		3N207	0
BSV38	NJD	B60	25V	150MA	150C	0.3WF	4/10V	50MNMA	25TPMO	25R	18P	8P	ALG	TIB			0
BSV38P	NJD	B61	25V	150MA	150C	0.1WF	4/10V	50MNMA	25TPMO	250R	18P	8P	ALG	TIB			0
BSV39	NJD	B60	25V	10MA	150C	0.3WF	0.8/6V	8/100MA	70TPMO		18P	8P	ALG	TIB			0
BSV39P	NJD	B61	25V	10MA	150C	0.1WF	0.8/6V	8/100MA	70TPMO		18P	8P	ALG	TIB			0
BSV78	NJD	B6	40V	50MA	175C	0.3WF	3.75/11V	50MNMA		25R	10P	5P	RLS	MUB	BSV78	2N4091	0
BSV79	NJD	B6	40V	50MA	175C	0.3WF	2/7V	20MNMA		40R	10P	5P	RLS	MUB	BSV78	2N4091	0
BSV80	NJD	B6	40V	50MA	175C	0.3WF	1/5V	10MNMA		60R	10P	5P	RLS	MUB	BSV78	2N4091	0
BSV81	NMD	B14	30V	50MA	125C	0.2WF	5MXV			50R	5P	OP5	RLS	MUB	BSV81	3N216	0

TYPE NO	CONS TRUC TION	CASE & LEAD	V DS MAX	I D MAX	T J MAX	P TOT MAX	VP OR VT	IDSS OR IDOM	G MO	R DS MAX	C ISS MAX	C RSS MAX	USE	SUPP LIER	EUR SUB	USA SUB	ISS
BSW30	PME	B53	30V	500MA	175C	0.4WF	6MXV		5/6.2MO	250R	14P		RLS	OBS	BSX83	3N161	0
BSW31	PME	B53	30V	500MA	175C	0.4WF	6MXV		5/6.2MO	125R	14P		RLS	OBS	BSW95	3N168	0
BSW95	PME	B53	30V	20MA	175C	0.3WF	6MXV	20MA	0.6MNMO	1K5	4P		RLS	OBS	BSX83	3N174	0
BSW95A	PME	B53	30V	20MA	175C	0.3WF	6MXV	20MA	0.6MNMO	1K5	4P		RLS	OBS	BSX83	3N174	0
BSX34	PME	B38	30V	0.2A	175C	0.6WF	6MXV		1.5MNMO		10P		DUA	OBS	BSX86	2N4066	0
BSX82	NJD	B14	30V	50MA	200C	0.2WF		30MXMA			5P	OP8	ALG	MUB	BFW61	2N3822	0
BSX83	PME	B52	30V	20MA	175C	0.3WF	6MXV	12MXMA	0.4/0.6MO	1K	4P5		RLS	OBS	BSX83	3N174	0
BSX84	PME	B52	30V	20MA	175C	0.3WF	6MXV	12MXMA	0.7/0.8MO	800R	4P5		RLS	OBS	BSX83	3N174	0
BSX85	PME		30V	200MA	175C	0.6WF	6MXV		1.5/2.3MO	300R	7P		DUA	OBS		2N4067	0
BSX86	PME		30V	200MA	175C	0.6WF	6MXV		2.5/3.1MO	125R	7P		DUA	OBS		2N4067	0
C21	NJD	B14	50V	5MA	175C	0.2WF	4MXV	.5/2.5MA	1.5/4.5MO		5P	2P	ALN	SEV	BFW12	2N3821	0
C2306	NJD	B51	30V	5MA	175C	0.3WF	1/4.5V	0.4MNMA	0.5TPMO				DUA	SEV	BFQ10	2N4082	0
C38	NJD	B47	50V	1MA	175C	0.2WF	1MXV	0.2/.6MA	0.6/1.8MO	2K5	6P	3P	ALC	SEV	BF800	2N3687	0
C413N	NJD	B47	15V	50MA	200C	0.4WF	3MXV	10MNMA	25MNMO		80P		PHT	OBS			0
C610	NJD	B22	40V		175C	.25WF	20MXV						RLS	OBS		2N4392	0
C611	NJD	B22	40V		175C	.25WF	20MXV						RLS	OBS		2N5360	0
C612	NJD	B22	40V		175C	.25WF	20MXV						RLS	OBS		2N5360	0
C613	NJD	B22	40V		175C	.25WF	20MXV						RLS	OBS		2N5360	0
C614	NJD	B22	40V		175C	.25WF	10MXV						RLS	OBS		2N5358	0
C615	NJD	B22	40V		175C	.25WF	10MXV						RLS	OBS		2N5360	0
C620	NJD	B22	20V		175C	.25WF							ALG	OBS		2N5358	0
C621	NJD	B22	20V		175C	.25WF							ALG	OBS		2N5358	0
C622	NJD	B22	20V		175C	.25WF							ALG	OBS		2N5358	0
C623	NJD	B22	20V		175C	.25WF							ALG	OBS		2N5358	0
C624	NJD	B22	20V		175C	.25WF							ALN	OBS		2N5358	0

TYPE NO	CONS TRUC TION	CASE & LEAD	V DS MAX	I D MAX	T J MAX	P TOT MAX	VP OR VT	IDSS OR IDOM	G MO	R DS MAX	C ISS MAX	C RSS MAX	USE	SUPP LIER	EUR SUB	USA SUB	ISS
C625	NJD	B22	20V		175C	.25WF							ALN	OBS		2N5358	0
C650	NJD	B22	20V		175C	.25WF							RLS	OBS		2N5358	0
C651	NJD	B22	20V		175C	.25WF							RLS	OBS		2N5358	0
C652	NJD	B22	20V		175C	.25WF							RLS	OBS		2N5358	0
C653	NJD	B22	20V		175C	.25WF							RLS	OBS		2N5358	0
C673	NJD	B23	40V			.25WF	10MXV	6TPMA	2.5TPMO				ALG	OBS		2N4341	0
C674	NJD	B47	40V				10MXV	6TPMA	2.5TPMO				ALG	OBS		2N4341	0
C680	NJD	B23	30V		200C	0.2WF		0.4MXMA	0.2/0.5MO		3P		FVG	OBS	BF800	2N3687	0
C680A	NJD	B23	30V		200C	0.2WF		0.4MXMA	0.2/0.5MO		3P		FVG	OBS	BF800	2N3687	0
C681	NJD	B47	30V		200C	0.2WF	2.5MXV	0.4MXMA	0.2/0.5MO		3P		ALG	CRY	BF800	2N3687	0
C681A	NJD	B47	30V		200C	0.2WF	2.5MXV	0.4MXMA	0.2/0.5MO		3P		ALG	CRY	BF800	2N3687	0
C682	NJD	B23	30V		200C	0.2WF	5MXV	1.6MXMA	0.4/1MO		3P		ALG	OBS	BF800	2N3686	0
C682A	NJD	B23	30V		200C	0.2WF	5MXV	1.6MXMA	0.4/1MO		3P		ALG	OBS	BF800	2N3686	0
C683	NJD	B47	30V		200C	0.2WF	5MXV	1.6MXMA	0.4/1MO		3P		ALG	CRY	BF800	2N3686	0
C683A	NJD	B47	30V		200C	0.2WF	5MXV	1.6MXMA	0.4/1MO		3P		ALG	CRY	BF800	2N3686	0
C684	NJD	B23	30V		200C	0.2WF	10MXV	6MXMA	0.6/1.5MO		3P		ALG	OBS	BF808	2N3685	0
C684A	NJD	B23	30V		200C	0.2WF	10MXV	6MXMA	0.6/1.5MO		3P		ALG	OBS	BF808	2N3685	0
C685	NJD	B47	30V		200C	0.2WF	10MXV	6MXMA	0.6/1.5MO		3P		ALG	CRY	BF808	2N3685	0
C685A	NJD	B47	30V		200C	0.2WF	10MXV	6MXMA	0.6/1.5MO		3P		ALG	CRY	BF808	2N3685	0
C6690	NJD	B47	30V	10MA	200C	0.3WF	10MXV			500R	5P		RLS	OBS	BF244	2N4220	0
C6691	NJD	B47	25V	10MA	200C	0.3WF	10MXV			500R	5P		RLS	OBS	BF244	2N4220	0
C6692	NJD	B47	25V	10MA	200C	0.3WF	6MXV			900R	5P		RLS	OBS	BF244	2N4220	0
C80	NJD	B6	50V	10MA	150C	0.3WF	30MXV		.05/.25MO				ALG	OBS	BF800	2N3687	0
C81	NJD	B6	50V	10MA	150C	0.4WF	18MXV		0.04/.2MO				ALG	OBS	BF800	2N3687	0
C82A	NJD	B6	50V	10MA	150C	0.2WF	5MXV		.03/0.2MO				ALG	OBS	BF800	2N3687	0

TYPE NO	CONS TRUC TION	CASE & LEAD	V DS MAX	I D MAX	T J MAX	P TOT MAX	VP OR VT	IDSS OR IDOM	G MO	R DS MAX	C ISS MAX	C RSS MAX	USE	SUPP LIER	EUR SUB	USA SUB	ISS
C82B	NJD	B6	50V	10MA	150C	0.2WF	10MXV		.03/0.2MO				ALG	OBS	BF800	2N3687	0
C83A	NJD	B6	50V	10MA	150C	0.2WF	10MXV		.03/0.2MO				ALG	OBS	BF800	2N3687	0
C83B	NJD	B6	50V	10MA	150C	0.2WF	15MXV		.03/0.2MO				ALG	OBS	BF800	2N3687	0
C84	NJD	B6	50V	10MA	150C	0.2WF	20MXV		.03/0.2MO				ALG	OBS	BF800	2N3687	0
C85	NJD	B6	50V	10MA	150C	0.2WF	25MXV		.03/0.2MO				ALG	OBS	BF800	2N3687	0
C91	NJD	B47	25V			0.3WF	15MXV	7MNMA			3P		RLS	OBS	BSV80	2N4858	0
C92	NJD	B47	40V	75MA	175C	0.2WF	7MXV	15MNMA		50R	16P	5P	RMS	SEV	BSV79	2N4092	0
C93	NJD	B47	40V	30MA	175C	0.2WF	3MXV	5/30MA		100R	10P	3P	RLS	SEV	BSV80	2N4393	0
C94	NJD	B14	50V	20MA	175C	0.2WF	5MXV	0.5/5MA	0.5/4MO		5P	2P	ALN	SEV	BF808	2N4868	0
C94A	NJD	B47	25V	5MA	200C	.35WF	5MXV	5MXMA	0.5/2MO		3P		ALG	OBS	BF808	2N5457	0
C94E	NJD	B1	25V	10MA	150C	0.2WF	5MXV	2/10MA	2MNMO		10P	6P	ALC	SEV	BC264	2N3684	0
C94EGX2	NJD	B1	25V	10MA	150C	0.2WF	5MXV	2/10MA	2MNMO		10P	6P	MPP	SEV	BFS21A	MMF1	0
C94ERX2	NJD	B1	25V	10MA	150C	0.2WF	5MXV	2/10MA	2MNMO		10P	6P	MPP	SEV	BFS21A	MMF1	0
C95	NJD	B47	25V	10MA	200C	.35WF	15MXV	7MXMA	0.3/1.5MO		3P		ALG	OBS	BC264A	2N5457	0
C95A	NJD	B47	25V	10MA	200C	.35WF	15MXV	7MXMA	0.3/1.5MO		3P		ALG	OBS	BC264B	2N5457	0
C95E	NJD	B1	25V	20MA	150C	0.2WF	8MXV	8/18MA	2MNMO		10P	6P	ALC	SEV	BF244	2N5459	0
C95EG	NJD	B1	25V	20MA	150C	0.2WF	8MXV	8/18MA	2MNMO		10P	6P	MPP	SEV	BFS21A	MMF1	0
C95ER	NJD	B1	25V	20MA	150C	0.2WF	8MXV	8/18MA	2MNMO		10P	6P	MPP	SEV	BFS21A	MMF1	0
C96E	NJD	B1	25V	20MA	150C	0.2WF	8MXV	2/18MA	2MNMO		10P	6P	ALG	SEV	BF244	2N3819	0
C96EGX2	NJD	B1	25V	20MA	150C	0.2WF	8MXV	2/18MA	2MNMO		10P	6P	MPP	SEV	BFS21A	MMF1	0
C97E	NJD	B1	50V	15MA	150C	0.2WF	1.4/3V	3/11MA	3/9MO		10P	6P	ALC	SEV	BFW56	2N3822	0
C97EG	NJD	B1	50V	15MA	150C	0.2WF	1.4/3V	3/11MA	3/9MO		10P	6P	MPP	SEV	BFS21A	MMF1	0
C97ER	NJD	B1	50V	15MA	150C	0.2WF	1.4/3V	3/11MA	3/9MO		10P	6P	MPP	SEV	BFS21A	MMF1	0
C98E	NJD	B1	50V	15MA	150C	0.2WF	4MXV	13MXMA	3.5/8.5MO		10P	6P	ALC	SEV	BF805	2N3822	0
C98EG	NJD	B1	50V	15MA	150C	0.2WF	4MXV	13MXMA	3.5/8.5MO		10P	6P	MPP	SEV	BFS21A	MMF1	0

TYPE NO	CONS TRUC TION	CASE & LEAD	V DS MAX	I D MAX	T J MAX	P TOT MAX	VP OR VT	IDSS OR IDOM	G MO	R DS MAX	C ISS MAX	C RSS MAX	USE	SUPP LIER	EUR SUB	USA SUB	ISS
C98ER	NJD	B1	50V	15MA	150C	0.2WF	4MXV	13MXMA	3.5/8.5MO		10P	6P	MPP	SEV	BFS21A	MMF1	0
CC641	NJD	B73	20V	400MA	200C	0.4WF	2.5MXV	30MNMA		100R	16P	8P	RMS	CRY			0
CC645	NJD	B73	30V	400MA	200C	0.4WF	5MXV	30MNMA		50R	16P	8P	RMS	CRY			0
CC647	NJD	B73	30V	400MA	200C	0.4WF	10MXV	30MNMA		30R	16P	8P	RMS	CRY			0
CC697	NJD	B73	25V	400MA	200C	0.4WF	3MXV	30MNMA		15R	64P	32P	RMS	CRY		CM697	0
CC4445	NJD	B73	25V	400MA	200C	0.4WF	10MXV	150MNMA		6R	64P	32P	RMS	CRY		2N4447	0
CC4446	NJD	B73	25V	400MA	200C	0.4WF	10MXV	100MNMA		10R	64P	32P	RMS	CRY		2N4446	0
CF24	NJD	B58	25V	40MA	125C	0.2WF	7MXV	40MXMA	2/9MO	500R	12P		RLS	OBS		2N3824	0
CF2386	NJD	B21	20V	10MA	200C	0.5WF	8MXV	9MXMA	1MNMO		50P		ALG	OBS	BFW12	2N5458	0
CFM13026	NJD	B6	40V	50MA	175C	1.8WC	7MXV	50MXMA	15/25MO		20P		ALG	OBS	BF815	2N6451	0
CM600	NJD	B47	10V	100MA	200C	0.3WF	7MXV	50TPMA	10/30MO	60R	6P5		RMS	OBS	BSV80	2N4858	0
CM601	NJD	B47	10V	100MA	200C	0.3WF	10MXV	50MXMA	10/30MO	60R	6P5		RMS	OBS	BSV80	2N4392	0
CM602	NJD	B47	30V	100MA	200C	0.3WF	10MXV	70MXMA	10/30MO	40R	6P5		RMS	OBS	BSV79	2N4092	0
CM603	NJD	B47	15V	100MA	200C	0.3WF	1.5MXV	100MXMA	20/60MO	30R	6P5		RMS	OBS	BSV78	2N4391	0
CM640	NJD	B47	20V	100MA	200C	0.3WF	2.2MXV	0.5MNMA	5MNMO	250R	5P		RMS	CRY	BSV80	2N4393	0
CM641	NJD	B47	20V	100MA	200C	0.3WF	2.2MXV	3MNMA	10MNMO	100R	5P		RMS	CRY	BSV80	2N4393	0
CM642	NJD	B47	20V	100MA	200C	0.3WF	3MXV	10MNMA	20MNMO	50R	5P		RMS	CRY	BSV79	2N4392	0
CM643	NJD	B47	20V	100MA	200C	0.3WF	5MXV	50MNMA	30MNMO	35R	5P		RMS	CRY	BSV78	2N4391	0
CM644	NJD	B47	30V	100MA	200C	0.3WF	3MXV	10MNMA	20MNMO	50R	5P		RLS	CRY	BSV79	2N4092	0
CM645	NJD	B47	30V	100MA	200C	0.3WF	5MXV	15MNMA	20MNMO	40R	5P		RMS	CRY	BSV79	2N4857	0
CM646	NJD	B47	30V	100MA	200C	0.3WF	7MXV	30MNMA	30MNMO	30R	5P		RMS	CRY	BSV78	2N4391	0
CM647	NJD	B47	30V	100MA	200C	0.3WF	10MXV	50MNMA	30MNMO	25R	5P		RMS	CRY	BSV78	2N4859	0
CM697	NJD	B47	25V	400MA	200C	0.4WF	0.5/3V	30MNMA		15R	20P		RLS	CRY		2N5434	0
CM800	NJD	B47	30V	400MA	200C	0.4WF	1/7V	30MNMA		30R	6P		RLS	CRY	BSV78	2N4091	0
CMX740	NJD	B47	30V	500MA	200C	0.4WF	10MXV	500MXMA		2R5			RMS	OBS		2N6568	0

TYPE NO	CONS TRUC TION	CASE & LEAD	V DS MAX	I D MAX	T J MAX	P TOT MAX	VP OR VT	IDSS OR IDOM	G MO	R DS MAX	C ISS MAX	C RSS MAX	USE	SUPP LIER	EUR SUB	USA SUB	ISS
CP600	NJD	B83	20V	180MA	200C	3.6WC	8MXV	180MXMA	10/30MO	60R	35P		AMP	OBS			0
CP601	NJD	B83	30V	180MA	200C	5.4WC	8MXV	180MXMA	10/30MO	60R	35P		AMP	OBS			0
CP602	NJD	B83	20V	300MA	200C	6WC	15MXV	300MXMA	20/60MO	40R	35P		AMP	OBS			0
CP603	NJD	B83	30V	300MA	200C	9WC	15MXV	300MXMA	20/60MO		35P		AMP	OBS			0
CP640	NJD	B23	20V	1A2	200C	2WC	2/10V	0.1/.8A	40/80MO		20P		VMP	CRY		2N5433	0
CP643	NJD	B47	30V	300MA	200C	2WC	2/7V	50/250MA	20/30MO		6P		VMP	CRY		2N4859	0
CP650	NJD	B23	25V	1A2	200C	8WC	2/10V	0.3/1.2A	0.1/.25MO	8R	25P		UMP	CRY		2N6568	0
CP651	NJD	B23	20V	600MA	200C	8WC	2/10V	0.1/.5A	.075/.2MO	14R	25P		UMP	CRY		2N5432	0
CP652	NJD	B23	20V	600MA	200C	8WC	2/10V	100MNMA	0.1TPMO	6R	25P		RMS	CRY		2N5432	0
CP653	NJD	B23	20V	600MA	200C	8WC	2/10V	60MNMA	0.06TPMO	12R	25P		RMS	CRY		2N5434	0
D1101	NJD	B6	25V	4MA	200C	0.4WF	10MXV	4MXMA	0.4/2MO	2K8	2P		ALG	OBS	BF808	2N3821	0
D1102	NJD	B6	25V	4MA	200C	0.4WF	5MXV	1MXMA	0.3/1MO	3K6	2P		ALG	OBS	BF800	2N3821	0
D1103	NJD	B6	25V	4MA	200C	0.4WF	2.5MXV	0.25MXMA	0.2/1MO	5K5	2P		ALG	OBS		2N3687A	0
D1177	NJD	B6	50V	4MA	200C	0.4WF	10MXV	4MXMA	0.4/2MO	2K8	2P		ALG	OBS	BFW13	2N3821	0
D1178	NJD	B6	50V	4MA	200C	0.4WF	5MXV	1MXMA	0.3/1MO	3K6	2P		ALG	OBS	BF800	2N3686	0
D1179	NJD	B6	50V	4MA	200C	0.4WF	2.5MXV	0.25MXMA	0.2/1MO	5K5	2P		ALG	OBS	BF800	2N3686	0
D1180	NJD	B6	50V	4MA	200C	0.4WF	10MXV	4MXMA	1/4MO	1K1	3P5		ALG	OBS	BF808	2N4868	0
D1181	NJD	B6	50V	4MA	200C	0.4WF	5MXV	1MXMA	.75/2.5MO	1K5	3P5		ALG	OBS	BF800	2N3686	0
D1182	NJD	B6	50V	4MA	200C	0.4WF	2.5MXV	0.25MXMA	0.5/2.5MO	2K2	3P5		ALG	OBS	BF800	2N3687	0
D1183	NJD	B6	50V	15MA	200C	0.4WF	8MXV	15MXMA	2.5/10MO		6P		ALG	OBS	BF810	2N4303	0
D1184	NJD	B6	50V	15MA	200C	0.4WF	4MXV	4MXMA	1.5/6MO		6P		ALG	OBS	BFW12	2N5457	0
D1185	NJD	B6	50V	15MA	200C	0.4WF	2MXV	1MXMA	0.8/4.5MO		6P		ALG	OBS	BFW13	2N3686	0
D1201	NJD	B6	25V	10MA	200C	0.4WF	10MXV	10MXMA	1/4MO		5P		ALG	OBS	BFW12	2N5459	0
D1202	NJD	B6	25V	10MA	200C	0.4WF	5MXV	5MXMA	0.6/2.5MO	1K1	5P		ALG	OBS	BFW12	2N5457	0
D1203	NJD	B6	25V	10MA	200C	0.4WF	2.5MXV	0.6MXMA	0.3/2.5MO		5P		ALG	OBS	BF800	2N3687	0

TYPE NO	CONS TRUC TION	CASE & LEAD	V DS MAX	I D MAX	T J MAX	P TOT MAX	VP OR VT	IDSS OR IDOM	G MO	R DS MAX	C ISS MAX	C RSS MAX	USE	SUPP LIER	EUR SUB	USA SUB	ISS
D1301	NJD	B6	25V	15MA	200C	0.4WF	8MXV	15MXMA	2.5/10MO		6P		ALG	OBS	BFW61	2N5459	0
D1302	NJD	B6	25V	15MA	200C	0.4WF	4MXV	4MXMA	1.5/6MO		6P		ALG	OBS	BFW12	2N5457	0
D1303	NJD	B6	25V	15MA	200C	0.4WF	2MXV	1MXMA	0.8/4.5MO		6P		ALG	OBS	BFW13	2N3686	0
D1420	NJD	B6	25V	15MA	200C	0.4WF	4MXV	5MXMA	1MNMO		5P		ALG	OBS	BF808	2N4868	0
D1421	NJD	B6	25V	15MA	200C	0.4WF	6MXV	10MXMA	2MNMO		6P		ALG	MOB	BC264	2N3684	0
D1422	NJD	B6	25V	15MA	200C	0.4WF	10MXV	15MXMA	1MNMO		5P		ALG	OBS	BF805	2N4869	0
DA102	NJD	B51	50V		175C	.75WF							DUA	OBS	BFQ10	2N4082	0
DA402	NJD	B51	50V		175C	.75WF							DUA	OBS	BFQ10	2N4082	0
DD07K	PME	B53	30V	50MA	175C	0.3WF	5.5MXV		4/5.4MO		6P		ALG	OBS		3N160	0
DD08K	PME	B53	30V	50MA	175C	0.3WF	5.5MXV		4/5.4MO		6P		ALG	OBS		3N160	0
DD09K	PME	B53	30V	30MA	175C	0.3WF	5.5MXV		2MNMO	4K	3P		ALG	OBS		3N160	0
DD10K	PME	B53	30V	50MA	175C	0.3WF	2.5MXV		8MXMO		5P		ALG	OBS		3N160	0
DD11K	PME	B53	30V	30MA	175C	0.3WF	6MXV		1.4MNMO		3P		ALG	OBS		3N163	0
DD12J	PME	B16	30V	100MA	175C	.45WF	5.5MXV		9/14MO		13P		ALG	OBS		3N161	0
DD13K	PME	B39	30V	100MA	175C	.45WF	5.5MXV		9/14MO		13P		ALG	OBS		3N161	0
DM01B	PME	B40	30V	30MA	175C	0.5WF	6MXV		0.5/2MO		2P		DUA	OBS		3N189	0
DM02B	PME	B40	30V	30MA	175C	0.5WF	6MXV		0.5/2MO		2P		DUA	OBS		3N189	0
DM03B	PME	B40	30V	30MA	175C	0.5WF	6MXV		0.5/2MO		2P		DUA	OBS		3N190	0
DM05A	PME	B42	30V	30MA	175C	0.5WF	5.5MXV		2/6MO		3P		DUA	OBS		2N4067	0
DM06A	PME	B42	30V	30MA	175C	0.5WF	5.5MXV		2/6MO		3P		DUA	OBS		2N4067	0
DN3066A	NJD	B6	50V	4MA	200C	0.4WF	10MXV	4MXMA	0.4/1MO		10P		ALG	OBS	BFW13	2N3821	0
DN3067A	NJD	B6	50V	4MA	200C	0.4WF	5MXV	1MXMA	0.3/1MO		10P		ALG	OBS	BF800	2N3687	0
DN3068A	NJD	B6	50V	4MA	200C	0.4WF	2.5MXV	0.25MXMA	0.2/1MO		10P		ALG	OBS	BF800	2N3687	0
DN3069A	NJD	B6	50V	10MA	200C	0.4WF	10MXV	10MXMA	1/2.5MO		15P		ALG	OBS	BF808	2N4868	0
DN3070A	NJD	B6	50V	10MA	200C	0.4WF	5MXV	2.5MXMA	.75/2.5MO		15P		ALG	OBS	BF808	2N3685	0

TYPE NO	CONS TRUC TION	CASE & LEAD	V DS MAX	I D MAX	T J MAX	P TOT MAX	VP OR VT	IDSS OR IDOM	G MO	R DS MAX	C ISS MAX	C RSS MAX	USE	SUPP LIER	EUR SUB	USA SUB	ISS
DN3071A	NJD	B6	50V	10MA	200C	0.4WF	2.5MXV	0.6MXMA	0.5/2.5MO		15P		ALG	OBS	BF800	2N3687	0
DN3365A	NJD	B6	40V	4MA	150C	0.3WF	12MXV	4MXMA	0.4/2MO		15P		ALG	OBS	BF808	2N3365	0
DN3366A	NJD	B6	40V	4MA	150C		6.5MXV	1MXMA	0.25/1MO		15P		ALG	OBS	BF800	2N3365	0
DN3367A	NJD	B6	40V	4MA	150C	0.3WF	2.2MXV	0.25MXMA	0.1/1MO		15P		ALG	OBS	BF800	2N3367	0
DN3368A	NJD	B6	40V	12MA	150C	0.3WF	12MXV	12MXMA	1/4MO		20P		ALG	OBS	BC264	2N3368	0
DN3369A	NJD	B6	40V	12MA	150C	0.3WF	6.5MXV	2.5MXMA	0.6/2.5MO		20P		ALG	OBS	BFW13	2N3369	0
DN3370A	NJD	B6	40V	12MA	150C	0.3WF	3.2MXV	0.6MXMA	0.3/2.5MO		20P		ALG	OBS	BFW13	2N3370	0
DN3436A	NJD	B6	50V	15MA	200C	0.3WF	9.8MXV	15MXMA	2.5/10MO		18P		ALG	OBS	BFW61	2N3436	0
DN3437A	NJD	B6	50V	15MA	200C	0.3WF	4.8MXV	4MXMA	1.5/6MO		18P		ALG	OBS	BFW12	2N3437	0
DN3438A	NJD	B6	50V	15MA	200C	0.3WF	2.3MXV	1MXMA	0.8/4.5MO		18P		ALG	OBS	BFW13	2N3438	0
DN3458A	NJD	B6	50V	15MA	200C	0.3WF	7.8MXV	15MXMA	2.5/10MO		18P		ALG	OBS	BF805	2N3458	0
DN3459A	NJD	B6	50V	15MA	200C	0.3WF	3.4MXV	4MXMA	1.5/6MO		18P		ALN	OBS	BC264A	2N3459	0
DN3460A	NJD	B6	50V	15MA	200C	0.3WF	1.8MXV	1MXMA	0.8/4.5MO		18P		ALN	OBS	BFW13	2N3460	0
DN3684	NJD	B73	50V	8MA	200C	.35WF	2/5V	2.5/8MA	2/3MO	600R	4P	1P2	RLS	FCB		2N3684	0
DN3685	NJD	B73	50V	3MA	200C	.35WF	1/3.5V	1/3MA	1.5/2.5MO	800R	4P	1P2	RLS	FCB		2N3685	0
DN3686	NJD	B73	50V	2MA	200C	.35WF	0.6/2V	.4/1.2MA	1/2MO	1K2	4P	1P2	RLS	FCB		2N3686	0
DN4342	PJD	B73	25V	12MA	200C	.25WF	1/5.5V	4/12MA	2/6MO	700R	20P	5P	RLG	FCB	BF320C	2N2609	0
DN4343	PJD	B73	25V	30MA	135C	.18WF	2/10V	10/30MA	4/8MO	350R	20P	5P	RLG	FCB	BFT11	2N4343	0
DN4391	NJD	B73	40V	150MA	200C	0.3WF	4/10V	50/150MA		30R	14P	3P5	RLS	FCB	BSV78	2N4391	0
DN4392	NJD	B73	40V	75MA	200C	0.3WF	2/5V	25/75MA		60R	14P	3P5	RLS	FCB	BSV80	2N4392	0
DN4393	NJD	B73	40V	30MA	200C	0.3WF	0.5/3V	5/30MA		100R	14P	3P5	RLS	FCB		2N4393	0
DN5033	PJD	B73	20V	4MA	200C	0.3WF	0.3/2.5V	.3/3.5MA	1/5MO	1K3	25P	7P	FVG	FCB	BF320A	2N5033	0
DN5484	NJD	B73	25V	5MA	150C	0.3WF	0.3/3V	1/5MA	2.5MNMO		5P	1P2	FVG	FCB	BF256A	2N5484	0
DN5485	NJD	B73	25V	10MA	150C	0.3WF	0.5/4V	4/10MA	3MNMO		5P	1P2	FVG	FCB	BF256B	2N5484	0
DN5486	NJD	B73	25V	20MA	150C	0.3WF	2/6V	8/20MA	3.5MNMO		5P	1P2	FVG	FCB	BF256C	2N5486	0

TYPE NO	CONS TRUC TION	CASE & LEAD	V DS MAX	I D MAX	T J MAX	P TOT MAX	VP OR VT	IDSS OR IDOM	G MO	R DS MAX	C ISS MAX	C RSS MAX	USE	SUPP LIER	EUR SUB	USA SUB	ISS
DNX1	NJD	B6	50V	6MA	200C	0.4WF	8MXV	6MXMA	0.3/1.5MO	2K8	2P		ALG	OBS	BF808	2N3687	0
DNX1A	NJD	B6	50V	6MA	200C	0.4WF	8MXV	6MXMA	0.3/1.5MO	2K8	2P		ALG	OBS	BF808	2N3687	0
DNX2	NJD	B6	50V	6MA	200C	0.4WF	4MXV	1MXMA	0.3/1MO	3K6	2P		ALG	OBS	BF800	2N3687	0
DNX2A	NJD	B6	50V	6MA	200C	0.4WF	4MXV	1MXMA	0.3/1MO	3K6	2P		ALG	OBS	BF800	2N3687	0
DNX3	NJD	B6	50V	6MA	200C	0.4WF	2MXV	0.25MXMA	0.2/0.7MO	5K5	2P		ALG	OBS		2N3687A	0
DNX3A	NJD	B6	50V	6MA	200C	0.4WF	2MXV	0.25MXMA	0.2/0.7MO	5K5	2P		ALG	OBS		2N3687A	0
DNX4	NJD	B6	50V	10MA	200C	0.4WF	8MXV	10MXMA	1/2.5MO		2P		ALG	OBS	BF808	2N4869	0
DNX4A	NJD	B6	50V	10MA	200C	0.4WF	8MXV	10MXMA	1/2.5MO		2P		ALG	OBS	BF808	2N4869	0
DNX5	NJD	B6	50V	10MA	200C	0.4WF	4MXV	2.5MXMA	0.7/2.5MO		2P		ALG	OBS	BF808	2N4868	0
DNX5A	NJD	B6	50V	10MA	200C	0.4WF	4MXV	2.5MXMA	0.7/2.5MO		2P		ALG	OBS	BF808	2N4868	0
DNX6	NJD	B6	50V	10MA	200C	0.4WF	2MXV	0.6MXMA	0.5/2.5MO		2P		ALG	OBS		2N3687	0
DNX6A	NJD	B6	50V	10MA	200C	0.4WF	2MXV	0.6MXMA	0.5/2.5MO		2P		ALG	OBS		2N3687	0
DNX7	NJD	B6	50V	15MA	200C	0.4WF	8MXV	15MXMA	2.5/10MO		6P		ALG	OBS	BFW61	2N4341	0
DNX7A	NJD	B6	50V	15MA	200C	0.4WF	8MXV	15MXMA	2.5/10MO		6P		ALG	OBS	BFW61	2N4341	0
DNX8	NJD	B6	50V	15MA	200C	0.4WF	4MXV	4MXMA	1.5/6MO		6P		ALG	OBS	BFW12	2N4340	0
DNX8A	NJD	B6	50V	15MA	200C	0.4WF	4MXV	4MXMA	1.5/6MO		6P		ALG	OBS	BFW12	2N4340	0
DNX9	NJD	B6	50V	15MA	200C	0.4WF	2MXV	1MXMA	0.8/4.5MO		6P		ALG	OBS	BF800	2N4339	0
DNX9A	NJD	B6	50V	15MA	200C	0.4WF	2MXV	1MXMA	0.8/4.5MO		6P		ALG	OBS	BF800	2N4339	0
DP146	PJD	B7	25V	1MA	200C	0.4WF	6MXV	.025MXMA	0.06MNMO				FVG	OBS		2N3574	0
DP147	PJD	B6	25V	1MA	200C	0.4WF	6MXV	.065MXMA	0.18MNMO				ALG	OBS		2N3575	0
DPT200	NME	B15	25V		150C				1.5TPMO				FVG	OBS		2N4351	0
DPT201	NMD	B15	25V		150C				1.5TPMO				FVG	OBS		3N128	0
E100	NJD	B10	30V	20MA	150C	.11WF	0.3/10V	0.2/20MA	0.5MNMO	3K	8P	3P	ALG	TDY	BF347	2N5457	0
E101	NJD	B10	30V	20MA	150C	.22WF	0.3/1.5V	0.2/1MA	0.5MNMO	3K	8P	3P	ALG	TDY	BFW13	2N4302	0
E102	NJD	B10	30V	20MA	150C	.23WF	0.8/4V	.9/4.5MA	1MNMO	1K2	8P	3P	ALG	TDY	BC264A	2N5457	0

TYPE NO	CONS TRUC TION	CASE & LEAD	V DS MAX	I D MAX	T J MAX	P TOT MAX	VP OR VT	IDSS OR IDOM	G MO	R DS MAX	C ISS MAX	C RSS MAX	USE	SUPP LIER	EUR SUB	USA SUB	ISS
E103	NJD	B10	30V	20MA	150C	.22WF	2/10V	4/20MA	1.5MNMO	650R	8P	3P	ALG	TDY	BF244	2N5459	0
E105	NJD	B10	30V	1A	125C	.35WF	4.5/10V	500MNMA		3R	70P		RMS	SIU		2N6568	0
E106	NJD	B10	30V	1A	125C	.35WF	2/6V	200MNMA		6R	70P		RMS	SIU		2N5432	0
E107	NJD	B10	30V	1A	125C	.35WF	0.5/4.5V	100MNMA		8R	70P		RMS	SIU		2N5432	0
E108	NJD	B10	25V	400MA	125C	.35WF	3/10V	80MNMA		8R	30P		RMS	SIU		2N5432	0
E109	NJD	B10	25V	400MA	125C	.35WF	2/6V	40MNMA		12R	30P		RMS	SIU		2N5434	0
E110	NJD	B10	25V	400MA	125C	.35WF	0.5/4V	10MNMA		18R	30P		RMS	SIU		2N5434	0
E111	NJD	B10	30V	100MA	125C	.35WF	3/10V	20MNMA		30R	10P		RLS	SIU	BSV78	2N4091	0
E112	NJD	B10	30V	100MA	125C	.35WF	1/5V	5MNMA		50R	10P		RLS	SIU	BSV79	2N4092	0
E113	NJD	B10	30V	100MA	125C	.35WF	0.5/3V	2MNMA		100R	10P		RLS	SIU	BSV80	2N4093	0
E114	NJD	B10	25V	100MA	125C	.35WF	3/10V	15MNMA		150R	4P		RLS	SIU		2N5555	0
E174	PJD	B10	30V	100MA	125C	.35WF	5/10V	20/100MA		85R	20P		RLS	SIU		2N5114	0
E175	PJD	B10	30V	100MA	125C	.35WF	3/6V	7/60MA		125R	20P		RLS	SIU		2N5115	0
E176	PJD	B10	30V	100MA	125C	.35WF	1/4V	2/25MA		250R	20P		RLS	SIU		2N5116	0
E177	PJD	B10	30V	100MA	125C	.35WF	.8/2.25V	1.5/20MA		300R	20P		RLS	SIU		2N3382	0
E201	NJD	B10	40V	20MA	125C	.35WF	0.3/1.5V	0.2/1MA	0.5MNMO		5P	2P	ALG	SIU	BF800	2N3686A	0
E202	NJD	B10	40V	20MA	125C	.35WF	0.8/4V	.9/4.5MA	1MNMO		5P	2P	ALG	SIU	BC264A	2N3685A	0
E203	NJD	B10	40V	10MA	125C	.35WF	2/10V	4/20MA	1.5MNMO		5P	2P	ALG	SIU	BFW11	2N3684A	0
E210	NJD	B10	25V	40MA	125C	.35WF	1/3V	2/15MA	4/12MO		5P	1P5	RLG	SIU	BFW61	2N5245	0
E211	NJD	B10	25V	40MA	125C	.35WF	2.5/4.5V	7/20MA	7/12MO		5P	1P5	RLG	SIU	BF810	2N5486	0
E212	NJD	B10	25V	40MA	125C	.35WF	4/6V	15/40MA	7/12MO		5P	1P5	RLG	SIU	BF348	2N5397	0
E230	NJD	B10	40V	10MA	125C	.35WF	1/3V	0.7/3MA	1/2.5MO		15P	2P	ALN	SIU	BFW13	2N3685A	0
E231	NJD	B10	40V	10MA	125C	.35WF	2/6V	2/6MA	1.5/3MO		15P	2P	ALN	SIU	BFW12	2NB684A	0
E232	NJD	B10	40V	10MA	125C	.35WF	4/6V	5/10MA	2.5/4MO		15P	2P	ALN	SIU	BC264D	2N3684A	0
E270	PJD	B11	30V	50MA	125C	.35WF	0.5/2V	2/15MA	6/15MO		20P	5P	RLG	SIU		2N3934A	0

TYPE NO	CONSTRUCTION	CASE & LEAD	V DS MAX	I D MAX	T J MAX	P TOT MAX	VP OR VT	IDSS OR IDOM	G MO	R DS MAX	C ISS MAX	C RSS MAX	USE	SUPPLIER	EUR SUB	USA SUB	ISS
E271	PJD	B11	30V	50MA	125C	.35WF	1.5/4.5V	6/50MA	8/18MO		20P	5P	RLG	SIU		2N5116	0
E300	NJD	B10	25V	30MA	125C	.35WF	1/6V	6/30MA	4.5/9MO		5P5	1P7	FVG	SIU	BF256B	2N5397	0
E304	NJD	B10	30V	15MA	125C	.35WF	2/6V	5/15MA	4.5/7.5MO		3P	0P8	TUG	SIU	BF256	2N4416	0
E305	NJD	B10	30V	15MA	125C	.35WF	0.5/3V	1/8MA	3MNMO		3P	0P8	TUG	SIU	BF256A	2N5246	0
E308	NJD	B10	25V	60MA	125C	.35WF	1/6.5V	12/60MA	8/20MO		7P5		TUG	SIU	BF348	U308	0
E309	NJD	B10	25V	60MA	125C	.35WF	1/4V	12/30MA	10/20MO		7P5		TUG	SIU	BF348	U309	0
E310	NJD	B10	25V	60MA	125C	.35WF	2/6.5V	24/60MA	8/18MO		7P5		TUG	SIU	BF348	U310	0
E311	NJD	B10	25V	30MA	125C	.35WF	4MXV	30MXMA	10/20MO		5P		TUG	NAT		U311	0
E312	NJD	B10	25V	60MA	125C	.35WF	6.5MXV	60MXMA	8/18MO		5P		TUG	NAT		U310	0
E400	NJD	B71	40V	5MA	125C	.35WF	1/4.5V	0.5/5MA	1/4MO		4P5	1P2	DUA	SIU	BFQ10	2N5452	0
E401	NJD	B71	40V	5MA	125C	.35WF	1/4.5V	0.5/5MA	1/4MO		4P5	1P2	DUA	SIU	BFQ10	2N5452	0
E402	NJD	B71	40V	5MA	125C	.35WF	1/4.5V	0.5/5MA	1/4MO		4P5	1P2	DUA	SIU	BFQ10	2N5452	0
E410	NJD	B71	40V	6MA	125C	.35WF	0.5/3.5V	0.5/6MA	1/4MO		4P5	1P2	DUA	SIU	BFQ10	2N5452	0
E411	NJD	B71	40V	6MA	125C	.35WF	0.5/3.5V	0.5/6MA	1/4MO		4P5	1P2	DUA	SIU	BFQ10	2N5452	0
E412	NJD	B71	40V	6MA	125C	.35WF	0.5/3.5V	0.5/6MA	1/4MO		4P5	1P2	DUA	SIU	BFQ10	2N5452	0
E420	N JD	B71	25V	30MA	125C	.35WF	1/6V	6/30MA	4.5/9MO		3P5	0P8	DUA	SIU		2N5911	0
E421	NJD	B71	25V	30MA	125C	.35WF	1/6V	6/30MA	4.5/9MO		3P5	0P8	DUA	SIU		2N5912	0
E430	NJD	B44	25V	60MA	125C	.35WF	1/4V	10/30MA	10/20MO		7P5		DUA	SIU		2N5911	0
E431	NJD	B44	25V	60MA	125C	.35WF	2/6V	24/60MA	10/20MO		7P5		DUA	SIU			0
ESM25	NJD	B51	30V	15MA	200C	.25WF	0.7/4.5V	0.5/10MA	1/6MO		6P	2P	DUA	THS	BFQ10	2N5561	0
ESM25A	NJD	B51	30V	15MA	200C	.25WF	0.7/4.5V	0.5/10MA	1/6MO		6P	2P	DUA	THS	BFQ10	2N5561	0
ESM4091	NJD	B36	30V	100MA	150C	0.3WF	5/10V	30MNMA		30R	40P	8P	RLS	THS	BSV78	2N4091	0
ESM4092	NJD	B36	30V	50MA	150C	0.3WF	2/7V	15MNMA		50R	40P	8P	RLS	THS	BSV79	2N4092	0
ESM4093	NJD	B36	30V	25MA	150C	0.3WF	1/5V	8MNMA		80R	40P	8P	RLS	THS	BSV80	2N4093	0
ESM4302	NJD	B36	30V	25MA	150C	0.3WF	4MXV	0.5/5MA	1MNMO		12P	6P	ALN	THS	BF808	2N4868A	0

TYPE NO	CONS TRUC TION	CASE & LEAD	V DS MAX	I D MAX	T J MAX	P TOT MAX	VP OR VT	IDSS OR IDOM	G MO	R DS MAX	C ISS MAX	C RSS MAX	USE	SUPP LIER	EUR SUB	USA SUB	ISS
ESM4303	NJD	B36	30V	25MA	150C	0.3WF	6MXV	4/10MA	2MNMO		12P	6P	ALN	THS	BFW11	2N3684A	0
ESM4304	NJD	B36	30V	25MA	150C	0.3WF	10MXV	0.5/15MA	1MNMO		12P	6P	ALN	THS	BF805	2N4868A	0
ESM4446	NJD	B6	25V	100MA	200C	0.4WF	3/10V	100MNMA		8R	50P	25P	RLS	THS		2N5432	0
ESM4448	NJD	B6	25V	100MA	200C	0.4WF	1/5V	50MNMA		12R	50P	25P	RLS	THS		2N5434	0
FE0654A	NJD	B10	25V	40MA	125C	0.2WF	8MXV	40MXMA	4.5/9MO	150R	20P		RLS	OBS	BSV80	2N4393	0
FE0654B	NJD	B10	25V	40MA	125C	0.2WF	4MXV	12MXMA	3.5/8MO	220R	20P		RLS	OBS	BSV80	2N3966	0
FE0654C	NJD	B10	25V	40MA	125C	0.2WF	2.5MXV	4MXMA	2/6MO	300R	20P		RLS	OBS	BSV80	2N3966	0
FE1600	NJD	B22	30V	200MA	150C	0.4WF	15MXV			50R			RLS	OBS	BSV79	2N4092	0
FE3819	NJD	B10	25V	20MA	125C	0.3WF	8MXV	2/20MA	2/6.5MO		8P	4P	FVG	TDY	BF244	2N3819	0
FE4302	NJD	B10	30V	15MA	145C	0.3WF	4MXV	5MXMA	1MNMO		6P		ALG	OBS	BC264A	2N4302	0
FE4303	NJD	B10	30V	15MA	145C	0.3WF	6MXV	10MXMA	2MNMO		6P		ALG	OBS	BC264	2N4303	0
FE4304	NJD	B10	30V	15MA	145C	0.3WF	1MXV	15MXMA	1MNMO		6P		ALG	OBS	BC264D	2N4304	0
FE5245	NJD	B58	30V	25MA	150C	.36WF	6MXV	5/15MA	4.5/7.5MO		4P5	1P	TUG	OBS	BF256	2N5245	0
FE5246	NJD	B58	30V	25MA	150C	.36WF	4MXV	7MXMA	3/6MO		4P5	1P	TUG	OBS	BF256A	2N5246	0
FE5247	NJD	B58	30V	25MA	150C	.36WF	8MXV	24MXMA	4.5/8MO		4P5	1P	TUG	OBS	BF256C	2N5247	0
FE5457	NJD	B10	25V	20MA	135C	.31WF	6MXV	5MXMA	1/5MO		7P		ALG	OBS	BC264B	2N5457	0
FE5458	NJD	B10	25V	20MA	135C	.31WF	7MXV	9MXMA	1.5/5.5MO		7P		ALG	OBS	BC264C	2N5458	0
FE5459	NJD	B10	25V	20MA	135C	.31WF	8MXV	16MXMA	2/6MO		7P		ALG	OBS	BC264D	2N5459	0
FE5484	NJD	B58	25V	20MA	135C	.31WF	3MXV	1/5MA	3/6MO		5P	1P	FVG	OBS	BFW61	2N5484	0
FE5485	NJD	B58	25V	20MA	135C	.31WF	4MXV	4/10MA	3.5/7MO		5P	1P	FVG	OBS	BFW61	2N5485	0
FE5486	NJD	B58	25V	20MA	135C	.31WF	7.5MXV	8/20MA	4/8MO		5P	1P	FVG	OBS	BFW61	2N5486	0
FF102	NJD	B15	15V	10MA	200C	0.4WF	1/6V	0.8/5MA	0.8MNMO		8P		PHT	CRY			0
FF108	NJD	B15	15V	10MA	200C	0.4WF	1/6V	0.8/5MA	0.8MNMO		8P		PHT	CRY			0
FF400	NJD	B15	15V	6MA	200C	0.1WF	7MXV	6MXMA	1.5MNMO		5P		ALG	OBS	BFW12	2N5457	0
FF409	NJD	B6	25V	100MA	200C	0.3WF	5MXV	35TPMA	8/25MO		6P5		PHT	OBS			0

TYPE NO	CONS TRUC TION	CASE & LEAD	V DS MAX	I D MAX	T J MAX	P TOT MAX	VP OR VT	IDSS OR IDOM	G MO	R DS MAX	C ISS MAX	C RSS MAX	USE	SUPP LIER	EUR SUB	USA SUB	ISS
FF411	NJD	B6	25V	100MA	200C	0.3WF	5MXV	35TPMA	8/25MO		6P5		PHT	OBS			0
FF412	NJD	B6	30V	50MA	200C	0.3WF	1/5V	5MNMA	8/25MO	100R	6P5		PHT	CRY			0
FF413	NJD	B6	30V	50MA	200C	0.3WF	1/5V	5MNMA	8/25MO	100R	6P5		PHT	CRY			0
FF600	NJD	B15	15V	100MA	200C	0.3WF	5MXV	25TPMA			35P		PHT	OBS			0
FF617	NJD	B15	15V	100MA	200C	0.3WF	5MXV	25TPMA			35P		PHT	OBS			0
FF626	NJD	B15	15V	50MA	200C	0.4WF	1/5V	8MNMA	8MNMO		20P		PHT	CRY			0
FF627	NJD	B15	15V	50MA	200C	0.4WF	1/5V	8MNMA	8MNMO		20P		PHT	CRY			0
FI100	NME	B53	30V		150C	1WC			0.3MNMO				RLS			2N4351	0
FI0049	PME	B38	30V	200MA	175C	0.3WF	6MXV	50MXMA	2MNMO	125R	5P		DUA	OBS		2N4067	0
FM1100	NJD	B43	35V	2MA	200C	0.5WF	3MXV	1.2MXMA	0.5/3MO		5P		DUA	NAB	BFQ10	2N4082	0
FM1100A	NJD	B43	35V	50MA	150C	0.5WF	3MXV	1.2MXMA	0.5/3MO		5P		DUA	NAB	BFQ10	2N5196	0
FM1101	NJD	B43	35V	2MA	200C	0.5WF	3MXV	1.2MXMA	0.5/3MO		5P		DUA	NAB	BFQ10	2N4082	0
FM1101A	NJD	B43	35V	50MA	150C	0.5WF	3MXV	1.2MXMA	0.5/3MO		5P		DUA	NAB	BFQ10	2N5196	0
FM1102	NJD	B43	35V	2MA	200C	0.5WF	3MXV	1.2MXMA	0.5/3MO		5P		DUA	NAB	BFQ10	2N4082	0
FM1102A	NJD	B43	35V	50MA	150C	0.5WF	3MXV	1.2MXMA	0.5/3MO		5P		DUA	NAB	BFQ10	2N5196	0
FM1103	NJD	B43	35V	2MA	200C	0.5WF	3MXV	1.2MXMA	0.5/3MO		5P		DUA	NAB	BFQ10	2N4082	0
FM1103A	NJD	B43	35V	50MA	150C	0.5WF	3MXV	1.2MXMA	0.5/3MO		5P		DUA	NAB	BFQ10	2N5196	0
FM1104	NJD	B43	35V	2MA	200C	0.5WF	3MXV	1.2MXMA	0.5/3MO		5P		DUA	NAB	BFQ10	2N4082	0
FM1104A	NJD	B43	35V	50MA	150C	0.5WF	3MXV	1.2MXMA	0.5/3MO		5P		DUA	NAB	BFQ10	2N5196	0
FM1105	NJD	B43	35V	10MA	200C	0.5WF	6MXV	10MXMA	1/6MO		5P		DUA	NAB	BFQ10	2N4082	0
FM1105A	NJD	B43	35V	50MA	150C	0.5WF	6MXV	10MXMA	1/6MO		5P		DUA	NAB	BFQ10	2N5196	0
FM1106	NJD	B43	35V	10MA	200C	0.5WF	6MXV	10MXMA	1/6MO		5P		DUA	NAB	BFQ10	2N4082	0
FM1106A	NJD	B43	35V	50MA	150C	0.5WF	6MXV	10MXMA	1/6MO		5P		DUA	NAB	BFQ10	2N5196	0
FM1107	NJD	B43	35V	10MA	200C	0.5WF	6MXV	10MXMA	1/6MO		5P		DUA	NAB	BFQ10	2N4082	0
FM1107A	NJD	B43	35V	50MA	150C	0.5WF	6MXV	10MXMA	1/6MO		5P		DUA	NAB	BFQ10	2N5196	0

TYPE NO	CONS TRUC TION	CASE & LEAD	V DS MAX	I D MAX	T J MAX	P TOT MAX	VP OR VT	IDSS OR IDOM	G MO	R DS MAX	C ISS MAX	C RSS MAX	USE	SUPP LIER	EUR SUB	USA SUB	ISS
FM1108	NJD	B43	35V	10MA	200C	0.5WF	6MXV	10MXMA	1/6MO		5P		DUA	NAB	BFQ10	2N4082	0
FM1108A	NJD	B43	35V	50MA	150C	0.5WF	6MXV	10MXMA	1/6MO		5P		DUA	NAB	BFQ10	2N5196	0
FM1109	NJD	B43	35V	10MA	200C	0.5WF	6MXV	10MXMA	1/6MO		5P		DUA	NAB	BFQ10	2N4082	0
FM1109A	NJD	B43	35V	50MA	150C	0.5WF	6MXV	10MXMA	1/6MO		5P		DUA	NAB	BFQ10	2N5196	0
FM1110	NJD	B43	25V	10MA	200C	0.5WF	10MXV	10MXMA	1/6MO		5P		DUA	NAB	BFQ10	2N4082	0
FM1110A	NJD	B43	35V	50MA	150C	0.5WF	10MXV	10MXMA	0.5/6MO		5P		DUA	NAB	BFQ10	2N5196	0
FM1111	NJD	B43	25V	10MA	200C	0.5WF	10MXV	10MXMA	1/6MO				DUA	NAB	BFQ10	2N4082	0
FM1111A	NJD	B43	35V	50MA	150C	0.5WF	10MXV	10MXMA	0.5/6MO		5P		DUA	NAB	BFQ10	2N5196	0
FM1112	NJD	B43	25V	10MA	200C	0.5WF	10MXV	10MXMA	1/6MO				DUA	NAB	BFQ10	2N4082	0
FM1200	NJD	B43	35V	3MA	200C	0.5WF	2MXV	2.5MXMA	0.8/4.5MO		8P		DUA	NAB	BFQ10	2N5452	0
FM1201	NJD	B43	35V	3MA	200C	0.5WF	2MXV	2.5MXMA	0.8/4.5MO		8P		DUA	NAB	BFQ10	2N5452	0
FM1202	NJD	B43	35V	3MA	200C	0.5WF	2MXV	2.5MXMA	0.8/4.5MO		8P		DUA	NAB	BFQ10	2N5452	0
FM1203	NJD	B43	35V	3MA	200C	0.5WF	2MXV	2.5MXMA	0.8/4.5MO		8P		DUA	NAB	BFQ10	2N5452	0
FM1204	NJD	B43	35V	3MA	200C	0.5WF	2MXV	2.5MXMA	0.8/4.5MO		8P		DUA	NAB	BFQ10	2N5452	0
FM1205	NJD	B43	35V	20MA	200C	0.5WF	7MXV	20MXMA	3/10MO		8P		DUA	NAB	BFQ10	2N5564	0
FM1206	NJD	B43	35V	20MA	200C	0.5WF	7MXV	20MXMA	3/10MO		8P		DUA	NAB	BFQ10	2N5564	0
FM1207	NJD	B43	35V	20MA	200C	0.5WF	7MXV	20MXMA	3/10MO		8P		DUA	NAB	BFQ10	2N5564	0
FM1208	NJD	B43	35V	20MA	200C	0.5WF	7MXV	20MXMA	3/10MO				DUA	NAB	BFQ10	2N5564	0
FM1209	NJD	B43	35V	20MA	200C	0.5WF	7MXV	20MXMA	3/10MO		8P		DUA	NAB	BFQ10	2N5564	0
FM1210	NJD	B43	25V	20MA	200C	0.5WF	7MXV	20MXMA	0.8/10MO		8P		DUA	NAB	BFQ10	2N5564	0
FM1211	NJD	B43	25V	20MA	200C	0.5WF	7MXV	20MXMA	0.8/10MO		8P		DUA	NAB	BFQ10	2N5564	0
FM3954	NJD	B43	50V	5MA	200C	0.5WF	4.5MXV	5MXMA	1/4MO		4P		DUA	NAB	BFQ10	2N5452	0
FM3954A	NJD	B43	50V	5MA	200C	0.5WF	4.5MXV	5MXMA	1/4MO		4P		DUA	NAB	BFQ10	2N5452	0
FM3955	NJD	B43	50V	5MA	200C	0.5WF	4.5MXV	5MXMA	1/4MO		4P		DUA	NAB	BFQ10	2N5452	0
FM3955A	NJD	B43	50V	5MA	200C	0.5WF	4.5MXV	5MXMA	1/4MO		4P		DUA	NAB	BFQ10	2N5452	0

TYPE NO	CONS TRUC TION	CASE & LEAD	V DS MAX	I D MAX	T J MAX	P TOT MAX	VP OR VT	IDSS OR IDOM	G MO	R DS MAX	C ISS MAX	C RSS MAX	USE	SUPP LIER	EUR SUB	USA SUB	ISS
FM3956	NJD	B43	50V	5MA	200C	0.5WF	4.5MXV	5MXMA	1/4MO		4P		DUA	NAB	BFQ10	2N5452	0
FM3957	NJD	B43	50V	5MA	200C	0.5WF	4.5MXV	5MXMA	1/4MO		4P		DUA	NAB	BFQ10	2N5452	0
FM3958	NJD	B43	50V	5MA	200C	0.5WF	4.5MXV	5MXMA	1/4MO		4P		DUA	NAB	BFQ10	2N5452	0
FN1024	PME	B80	30V		150C	0.3WF	3MXV		1/2.4MO	800R	2P		RLS	OBS		3N156	0
FN1034	PME	B80	15V		150C	0.3WF	3MXV		1/2.4MO	500R	2P		RLS	OBS		3N156	0
FP4339	PJD	B15	40V	4MA	175C	0.3WF	1.8MXV	1.5MXMA	0.8/2.4MO	1K7	7P		ALG	OBS	BF320A	2N5265	0
FP4340	PJD	B15	40V	4MA	175C	0.3WF	3MXV	3.6MXMA	1.3/3MO	1K5	7P		ALG	OBS	BF320A	2N5267	0
FT57	NMD	B52	15V	30MA	175C	.38WF		26MXMA	6MNMO		2P7		ALG		BFX63	2N3631	0
FT704	PME	B38	30V		175C	.35WF			25/100MO	1K5	4P5		DUA	OBS		3N188	0
FT3820	PJD	B50	20V	10MA	125C	0.2WF	8MXV	15MXMA	0.8/5MO		32P		ALG	OBS	BF320	2N3909	0
FT3909	PJD	B7	20V	25MA	125C	0.3WF	8MXV	15MXMA	1/5MO		32P		ALG	OBS	BF320	2N3909	0
FT0654A	NJD	B6	50V	40MA	175C	0.3WF	8MXV	40MXMA	4.5/9MO	150R	20P		RLS	OBS		2N5555	0
FT0654B	NJD	B6	50V	40MA	175C	0.3WF	8MXV	40MXMA	4.5/9MO	150R	20P		RLS	OBS		2N5555	0
FT0654C	NJD	B6	50V	40MA	175C	0.3WF	4MXV	12MXMA	3.5/8MO	220R	20P		RLS	OBS		2N3966	0
FT0654D	NJD	B6	50V	40MA	175C	0.3WF	4MXV	12MXMA	3.5/8MO	220R	20P		RLS	OBS		2N3966	0
FT0654E	NJD	B6	50V	40MA	175C	0.3WF	2.5MXV	4MXMA	2/6MO	300R	20P		RLS	OBS		2N3966	0
HA2000	PMD	B12	30V		175C	.35WF			1/2MO					OBS			0
HA2001	PMD	B12	35V		175C	.35WF			1/2MO					OBS			0
HA2010	PMD	B12	35V		175C	.35WF			1/2MO					OBS			0
HA2020	PMD	B12	35V		175C	.35WF			1/2MO					OBS			0
HEPF0021	NJD	B3	25V	30MA	150C	.31WF		20MXMA	3.5TPMO				ALG	MOB	BF244	2N3819	0
HEPF0021RT	NJD	B3	25V	30MA	150C	0.3WF		20MXMA	3.5MNMO				FVG	MOU	BFW61	2N5485	0
HEPF1035	PJD	B63	40V	20MA	135C	0.3WF	8MXV	14MXMA	1MNMO		7P		ALG	OBS	BF320B	2N5464	0
HEPF1035RT	PJD	B63	40V	20MA	135C			14MXMA	1MNMO				ALG	MOU	BF320	2N4342	0
HEPF2004	NJD	B18		50MA	175C	0.4WF		30MXMA	6TPMO				ALG	MOB	BF244	2N3819	0

TYPE NO	CONS TRUC TION	CASE & LEAD	V DS MAX	I D MAX	T J MAX	P TOT MAX	VP OR VT	IDSS OR IDOM	G MO	R DS MAX	C ISS MAX	C RSS MAX	USE	SUPP LIER	EUR SUB	USA SUB	ISS
HEPF2005	NJD	B14	20V	10MA	175C	0.3WF		10MXMA	2TPMO				ALG	MOB	BC264	2N4303	0
HEPF2007	NJD	B84	20V	30MA	175C	0.5WF	4MXV	30MXMA	10TPMO		6P		ALG	MOB		MPF121	0
HEPF2004RT	NMD	B18	20V	50MA	175J	0.4WF		30MXMA	6MNMO				FVG	MOU		3N140	0
HEPF2005RT	NJD	B14	30V	10MA	175C	0.3WF		10MXMA	2MNMO				ALG	MOU	BC264	2N3684A	0
HEPF2007RT	NMD	B84	25V	30MA	175C	0.5WF		30MXMA	10MNMO				TUG	MOU	BF905	3N209	0
HEP801	NJD	B14	20V	15MA	125C	0.2WF		9MXMA	3MNMO				ALG	MOB	BFW61	2N3822	0
HEP801RT	NJD	B14	30V	15MA	125C	0.2WF		9MXMA	3MNMO				FVG	MOU	BF256A	2N5485	0
HEP802	NJD	B3	25V	20MA	125C	0.2WF		20MXMA	2MNMO				ALG	MOB	BF244	2N3819	0
HEP802RT	NJD	B3	25V	20MA	125C	0.2WF		20MXMA	2MNMO				FVG	MOU	BFW61	2N4224	0
HEP803	PJD	B12	40V	20MA	175C	0.2WF		7MXMA	1MNMO				ALG	OBS	BF320A	2N3330	0
HEP803RT	PJD	B12	40V	20MA	175C	0.2WF		7MXMA	1MNMO				ALG	MOU	BF320B	2N3332	0
HR3N187	N	B18	20V	50MA	175C	.33WF	0.5/4V	5/30MA	7/18MO		8P5	0P03	TUG	RCU		3N187	0
HR3N200	NMD	B18	20V	50MA	175C	.33WF	0.1/3V	0.5/12MA	10/20MO		8P5	0P03	TUG	RCU		3N200	0
IMF3954	NJD	B51	40V	5MA	200C	.25WF	4.5MXV	0.5/5MA	1MNMO				DUA	TDY		2N5196	0
IMF3954A	NJD	B51	40V	5MA	200C	.25WF	4.5MXV	0.5/5MA	1MNMO				DUA	TDY		2N5515	0
IMF3955	NJD	B51	40V	5MA	200C	.25WF	4.5MXV	0.5/5MA	1MNMO				DUA	TDY		2N3955	0
IMF3955A	NJD	B51	40V	5MA	200C	.25WF	4.5MXV	0.5/5MA	1MNMO				DUA	TDY		2N3955A	0
IMF3956	NJD	B51	40V	5MA	200C	.25WF	4.5MXV	0.5/5MA	1MNMO				DUA	TDY		2N3956	0
IMF3957	NJD	B51	40V	5MA	200C	.25WF	4.5MXV	0.5/5MA	1MNMO				DUA	TDY		2N3957	0
IMF3958	NJD	B51	40V	5MA	200C	.25WF	4.5MXV	0.5/5MA	1MNMO				DUA	TDY		2N3958	0
IMF6485	NJD	B51	50V	7.5MA	150C	0.5WF	0.7/4V	.5/7.5MA	1/4MO		20P	3P5	DUA	INB	BFQ10	2N5515	0
IT100	PJD	B7	35V	100MA	200C	0.3WF	2/4.5V	10MNMA	8MNMO	75R	35P	12P	RLS	INB		2N5114	0
IT100/D	PJD	B73	35V	100MA	200C	0.3WF	2/4.5V	10MNMA	8MNMO	75R	35P	12P	RLS	INB		2N5114	0
IT100/W	PJD	B74	35V	100MA	200C	0.3WF	2/4.5V	10MNMA	8MNMO	75R	35P	12P	RLS	INB		2N5114	0
IT100T092	PJD	B63	35V	100MA	125C	0.3WF	2/4.5V	10MNMA	8MNMO	75R	35P	12P	RLS	INB		2N5114	0

TYPE NO	CONSTRUCTION	CASE & LEAD	V DS MAX	I D MAX	T J MAX	P TOT MAX	VP OR VT	IDSS OR IDOM	G MO	R DS MAX	C ISS MAX	C RSS MAX	USE	SUPPLIER	EUR SUB	USA SUB	ISS
IT101	PJD	B7	35V	100MA	200C	0.3WF	4/10V	20MNMA	8MNMO	60R	35P	12P	RLS	INB		2N5114	0
IT101/D	PJD	B73	35V	100MA	200C	0.3WF	4/10V	20MNMA	8MNMO	60R	35P	12P	RLS	INB		2N5114	0
IT101/W	PJD	B74	35V	100MA	200C	0.3WF	4/10V	20MNMA	8MNMO	60R	35P	12P	RLS	INB		2N5114	0
IT101T092	PJD	B63	35V	100MA	125C	0.3WF	4/10V	20MNMA	8MNMO	60R	35P	12P	RLS	INB		2N5114	0
IT108	NJD	B10	25V	25MA	125C	0.2WF	6MXV	25MXMA	4/8MO		5P		ALG	OBS	BF244	2N5486	0
IT200	NJD	B73	50V	30MA	200C	.12WH	6MXV	30MXMA	6MNMO	150R	7P		RLS	OBS			0
IT210	NJD	B73	50V	30MA	200C	.12WH	4MXV	12MXMA	6.5MNMO	250R	7P		RLS	OBS			0
IT220	NJD	B73	50V	30MA	200C	.12WH	2.5MXV	8MXMA	3MNMO	350R	7P		RLS	OBS			0
IT400	NJD	B6	40V	150MA	200C	0.3WF	4/10V	50/150MA		30R			RLS	INB			0
IT400/D	NJD	B73	40V	150MA	200C		4/10V	50/150MA		30R			RLS	INB			0
IT400/W	NJD	B74	40V	150MA	200C		4/10V	50/150MA		30R			RLS	INB			0
IT400T092	NJD	B63	40V	150MA	125C	0.3WF	4/10V	50/150MA		30R			RLS	INB			0
IT1700	PME	B56	40V	50MA	150C	.37WF	2/5V	2MNMA		400R	5P	1P2	ALN	INB		2N4352	0
IT1700/D	PME	B73	40V	50MA	150C	.37WF	2/5V	2MNMA		400R	5P	1P2	ALN	INB		2N4352	0
IT1700/W	PME	B74	40V	50MA	150C	.37WF	2/5V	2MNMA		400R	5P	1P2	ALN	INB		2N4352	0
IT1750	NME	B56	25V	100MA	150C	.37WF	0.5/3V	10MNMA	3MNMO	50R	6P	1P6	RLS	INB			0
IT1750/D	NME	B73	25V	100MA	150C	.37WF	0.5/3V	10MNMA	3MNMO	50R	6P	1P6	RLS	INB			0
IT1750/W	NME	B74	25V	100MA	150C	.37WF	0.5/3V	10MNMA	3MNMO	50R	6P	1P6	RLS	INB			0
ITE3066	NJD	B10	30V	5MA	125C	0.2WF	10MXV	4MXMA	0.3MNMO		10P		ALG	OBS	BC264A	2N3685	0
ITE3067	NJD	B10	30V	5MA	125C	0.2WF	5MXV	1MXMA	0.25MNMO		10P		ALG	OBS	BF800	2N3067	0
ITE3068	NJD	B10	30V	5MA	125C	0.2WF	2.5MXV	0.25MXMA	0.15MNMO		10P		ALG	OBS	BF800	2N3068	0
ITE4117	NJD	B10	40V	1MA	125C	0.2WF	1.8MXV	0.09MXMA	0.06MNMO		3P		ALN	OBS	BF800	2N3687	0
ITE4118	NJD	B10	40V	1MA	125C	0.2WF	3MXV	0.24MXMA	0.07MNMO		3P		FVG	OBS	BF800	2N3687	0
ITE4119	NJD	B10	40V	1MA	125C	0.2WF	6MXV	1MXMA	0.09MNMO		3P		FVG	OBS	BF800	2N3686	0
ITE4338	NJD	B10	40V	1MA	125C	0.2WF	1.5MXV	0.2/.6MA	0.5MNMO	2K5	7P	3P	ALG	TDY	BF800	2N4338	0

TYPE NO	CONS TRUC TION	CASE & LEAD	V DS MAX	I D MAX	T J MAX	P TOT MAX	VP OR VT	IDSS OR IDOM	G MO	R DS MAX	C ISS MAX	C RSS MAX	USE	SUPP LIER	EUR SUB	USA SUB	ISS
ITE4339	NJD	B10	40V	2MA	125C	0.2WF	2.5MXV	.5/1.5MA	0.7MNMO	1K7	7P	3P	ALG	TDY	BF347	2N4339	0
ITE4340	NJD	B6	40V	4MA	125C	0.2WF	3.5MXV	1.2/4MA	1MNMO	1K5	7P	3P	ALG	TDY	BF808	2N4340	0
ITE4341	NJD	B10	40V	10MA	125C	0.2WF	7MXV	3/9MA	1.5MNMO	800R	7P	3P	ALG	TDY	BC264B	2N4341	0
ITE4391	NJD	B10	30V	150MA	125C	0.3WF	10MXV	50/150MA		35R	16P	5P	RLS	TDY	BSV78	2N4391	0
ITE4392	NJD	B10	30V	75MA	125C	0.3WF	5MXV	75MA		65R	16P	5P	RLS	TDY	BSV80	2N4392	0
ITE4393	NJD	B10	30V	30MA	125C	0.3WF	3MXV	5/30MA		110R	16P	5P	RLS	TDY	BSV80	2N4393	0
ITE4416	NJD	B10	25V	20MA	125C	0.2WF	7MXV	4/20MA	3MNMO		5P	1P2	TUG	TDY	BF255	2N4416	0
ITE4867	NJD	B10	35V	2MA	125C	0.2WF	2MXV	1.2MXMA	0.7MNMO		25P	5P	ALN	TDY	BF800	2N4867	0
ITE4868	NJD	B10	35V	3MA	125C	0.2WF	3MXV	1/3MA	1MNMO		25P	5P	ALN	TDY	BF808	2N4868	0
ITE4869	NJD	B10	35V	8MA	125C	0.2WF	5MXV	2.5/8MA	1.3MNMO		25P	5P	ALN	TDY	BFW56	2N4869	0
J2N3823	NJD	B73	30V	20MA	200C	0.3WH	8MXV	4/20MA	3.5/6.5MO		6P	2P	FVG	THS	BFW61	2N3823	0
J2N3966	NJD	B73	30V	20MA	200C	0.3WH	4/6V	2/20MA		220R	6P	2P	RLS	THS		2N3966	0
J2N4091	NJD	B73	40V	150MA	175C	1.8WC	5/10V	30MNMA		30R	16P	5P	RLS	THS	BSV78	2N4091	0
J2N4092	NJD	B73	40V	15MA	175C	1.8WC	2/7V	15MNMA		50R	16P	5P	RLS	THS	BSV79	2N4092	0
J2N4093	NJD	B73	40V	50MA	175C	1.8WC	1/5V	8MNMA		80R	16P	5P	RLS	THS	BSV80	2N4093	0
J2N4220	NJD	B73	30V	15MA	200C	0.3WH	4MXV	0.5/3MA	1/4MO		6P	2P	FVG	THS	BF347	2N4220	0
J2N4221	NJD	B73	30V	15MA	200C	0.3WH	6MXV	2/6MA	2/5MO		6P	2P	FVG	THS	BFW12	2N4221	0
J2N4222	NJD	B73	30V	15MA	200C	0.3WH	8MXV	5/15MA	2.5/6MO		6P	2P	FVG	THS	BFW61	2N4222	0
J2N4391	NJD	B73	40V	150MA	175C	1.8WC	4/10V	50/150MA		30R	14P	4P	RLS	THS	BSV78	2N4391	0
J2N4392	NJD	B73	40V	75MA	175C	1.8WC	2/5V	25/75MA		60R	14P	4P	RLS	THS	BSV79	2N4392	0
J2N4393	NJD	B73	40V	30MA	175C	1.8WC	0.5/3V	5/30MA		100R	14P	4P	RLS	THS	BSV80	2N4393	0
J2N4416	NJD	B73	30V	15MA	200C	0.3WH	6MXV	5/15MA	4.5/7.5MO		4P	0P9	TUG	THS	BF256	2N4416	0
J308	NJD	B3	25V	60MA	125C	.35WF	1/6.5V	12/60MA	8/20MO		7P5		TUG	SIU	BF348	2N5398	0
J309	NJD	B3	25V	60MA	125C	.35WF	1/4V	12/30MA	10/20MO		7P5		TUG	SIU	BF348	2N5398	0
J310	NJD	B3	25V	60MA	125C	.35WF	2/6.5V	24/60MA	8/18MO		7P5		TUG	SIU	BF348	2N5397	0

TYPE NO	CONS TRUC TION	CASE & LEAD	V DS MAX	I D MAX	T J MAX	P TOT MAX	VP OR VT	IDSS OR IDOM	G MO	R DS MAX	C ISS MAX	C RSS MAX	USE	SUPP LIER	EUR SUB	USA SUB	ISS
JH2101	NJD	B51	50V	10MA	200C	0.3WF	5MXV	10MXMA	2/7MO		20P		DUA	OBS	BFQ10	2N5561	0
JH2102	NJD	B51	50V	10MA	200C	0.3WF	5MXV	10MXMA	2/7MO		20P		DUA	OBS	BFQ10	2N5561	0
JH2103	NJD	B51	50V	10MA	200C	0.3WF	5MXV	10MXMA	2/7MO		20P		DUA	OBS	BFQ10	2N5561	0
JH2104	NJD	B51	50V	10MA	200C	0.3WF	5MXV	10MXMA	2/7MO		20P		DUA	OBS	BFQ10	2N5561	0
JH2105	NJD	B51	50V	10MA	200C	0.3WF	5MXV	10MXMA	2/7MO		20P		DUA	OBS	BFQ10	2N5561	0
JH2106	NJD	B51	50V	10MA	200C	0.3WF	5MXV	10MXMA	2/7MO		20P		DUA	OBS	BFQ10	2N5561	0
K1001	NMD	B56	15V	40MA	175C	.15WF	6MXV	12MXMA	2/5MO		3P		FVG	OBS		3N128	0
K1002	NMD	B56	15V	40MA	175C	.15WF	6MXV	5MXMA	1/5MO		3P		FVG	OBS		3N128	0
K1003	NMD	B56	15V	40MA	175C	.15WF	6MXV	20MXMA	2.5TPMO		3P		FVG	OBS		3N128	0
K1004	NMD	B56	15V	40MA	175C	.15WF	12MXV	7MXMA	0.8/2.5MO		2P		FVG	OBS		3N128	0
K1201	NMD	B56	15V	15MA	175C	75MWF	5MXV	5MXMA	1/2MO				TUG	OBS	BFX63	2N3631	0
K1202	NMD	B56	15V	15MA	175C	75MWF	5MXV	5MXMA	1/2MO				TUG	OBS	BFX63	2N3631	0
K1501	PME	B56	15V	35MA	175C	.15WF	7MXV		1/2MO	300R	1P		RLS	OBS		3N155A	0
K1502	PME	B56	15V	35MA	175C	.15WF	7MXV		1/2MO	300R	1P		RLS	OBS		3N155A	0
K1504	PME	B56	15V	35MA	175C	.15WF	8MXV		0.8/2MO	700R	1P		RLS	OBS		3N155	0
KE510	NJD	B10	30V	50MA	125C	0.3WF	0.5/10V	5/50MA	8TPMO	100R	18P	6P	RLS	NAT	BSV80	2N4861	0
KE511	NJD	B10	20V	50MA	125C	0.3WF	10MXV	5/50MA	8TPMO	100R	18P	8P	RLS	NAT	BSV80	2N4861	0
KE3684	NJD	B10	50V	8MA	125C	.36WF	2/5V	2.5/7MA	2/3MO	600R	4P	1P2	ALN	NAB	BC264B	2N3684A	0
KE3685	NJD	B10	50V	8MA	125C	.36WF	1/3.5V	1/3MA	1.5/2.5MO	800R	4P	1P2	ALN	NAB	BC264A	2N3685A	0
KE3686	NJD	B10	50V	8MA	125C	.36WF	0.6/2V	.4/1.2MA	1/2MO	1K2	4P	1P2	ALN	NAB	BF808	2N3686A	0
KE3687	NJD	B10	50V	8MA	125C	.36WF	0.3/1.2V	0.1/.5MA	0.5/1.5MO		4P	1P2	ALN	NAB	BF800	2N3687A	0
KE3823	NJD	B10	30V	20MA	125C	0.3WF	8MXV	4/20MA	3.5/6.5MO		6P	2P	FVG	NAT	BF256B	2N3823	0
KE3970	NJD	B10	40V	150MA	125C	.25WF	4/10V	50/150MA		30R	25P	6P	RLS	TDY	BSV78	2N4391	0
KE3971	NJD	B10	40V	75MA	125C	.25WF	2/5V	75MA		60R	25P	6P	RLS	TDY	BSV79	2N4392	0
KE3972	NJD	B10	40V	30MA	125C	.25WF	0.5/3V	5/30MA		100R	25P	6P	RLS	TDY	BSV80	2N4393	0

TYPE NO	CONS TRUC TION	CASE & LEAD	V DS MAX	I D MAX	T J MAX	P TOT MAX	VP OR VT	IDSS OR IDOM	G MO	R DS MAX	C ISS MAX	C RSS MAX	USE	SUPP LIER	EUR SUB	USA SUB	ISS
KE4091	NJD	B10	40V	150MA	125C	.25WF	5/10V	30MNMA		30R	16P	5P	RLS	TDY	BSV78	2N4091	0
KE4092	NJD	B10	40V	100MA	125C	.25WF	2/7V	15MNMA		50R	16P	5P	RLS	TDY	BSV79	2N4092	0
KE4093	NJD	B10	40V	80MA	125C	.25WF	1/5V	8MNMA		80R	16P	5P	RLS	TDY	BSV80	2N4093	0
KE4220	NJD	B10	30V	20MA	125C	.25WF	4MXV	0.5/3MA	1/4MO		6P	2P	ALG	TDY	BF347	2N4220	0
KE4221	NJD	B10	30V	18MA	125C	.25WF	6MXV	2/6MA	2/5MO		6P	2P	ALG	TDY	BFW12	2N4221	0
KE4222	NJD	B10	30V	18MA	125C	.25WF	8MXV	5/15MA	2.5/6MO		6P	3P	ALG	TDY	BF244	2N4222	0
KE4223	NJD	B10	30V	20MA	125C	.25WF	0.1/8V	3/18MA	3/7MO		6P	2P	ALG	TDY	BF244	2N4223	0
KE4224	NJD	B10	30V	20MA	125C	.25WF	0.1/8V	2/20MA	2/7.5MO		6P	2P	ALG	TDY	BFW61	2N4224	0
KE4391	NJD	B10	40V	150MA	125C	.25WF	4/10V	50/150MA		30R	14P	3P5	RLS	TDY	BSV78	2N4391	0
KE4392	NJD	B10	40V	75MA	125C	.25WF	2/5V	25/75MA		60R	14P	3P5	RLS	TDY	BSV79	2N4392	0
KE4393	NJD	B10	40V	30MA	125C	.25WF	0.5/3V	5/30MA		100R	14P	3P5	RLS	TDY	BSV80	2N4393	0
KE4416	NJD	B10	30V	15MA	125C	.25WF	6MXV	5/15MA	4MNMO		4P	1P2	TUG	TDY	BF256	2N4416	0
KE4856	NJD	B10	40V	200MA	125C	.25WF	4/10V	50MNMA		25R	18P	8P	RLS	TDY	BSV78	2N4856	0
KE4857	NJD	B10	40V	100MA	125C	.25WF	2/6V	20/100MA		40R	18P	8P	RLS	TDY	BSV79	2N4857	0
KE4858	NJD	B10	40V	80MA	125C	.25WF	0.8/4V	8/80MA		60R	18P	8P	RLS	TDY	BSV80	2N4858	0
KE4859	NJD	B10	30V	200MA	125C	.25WC	4/10V	50MNMA		25R	18P	8P	RLS	TDY	BSV78	2N4859	0
KE4860	NJD	B10	30V	100MA	125C	.25WF	2/6V	20/100MA		40R	18P	8P	RLS	TDY	BSV79	2N4860	0
KE4861	NJD	B10	30V	80MA	125C	.25WF	0.8/4V	8/80MA		60R	18P	8P	RLS	TDY	BSV80	2N4861	0
KE5103	NJD	B10	25V	8MA	125C	.25WF	0.5/4V	1/8MA	2/8MO		5P	1P2	TUG	TDY	BF256A	2N5246	0
KE5104	NJD	B10	25V	6MA	125C	.25WF	0.5/4V	2/6MA	3.5/7.5MO		5P	1P2	TUG	TDY	BF256A	2N5104	0
KE5105	NJD	B10	25V	15MA	125C	.25WF	0.5/4V	5/15MA	5/10MO		5P	1P2	TUG	TDY	BF256	2N4416	0
LDF603	NJD	B60	20V	30MA	150C	.36WH	8MXV	8MXMA	1/4MO		5P		ALG	OBS			0
LDF604	NJD	B60	20V	30MA	150C	.36WH	8MXV	12MXMA	2.5/6.5MO		5P		ALG	OBS			0
LDF605	NJD	B60	20V	30MA	150C	.36WH	8MXV	20MXMA	3/7MO		5P		ALG	OBS			0
LDF691	NJD	B60	30V	50MA	150C	.36WH	10MXV			30R	16P		RLS	OBS			0

TYPE NO	CONS TRUC TION	CASE & LEAD	V DS MAX	I D MAX	T J MAX	P TOT MAX	VP OR VT	IDSS OR IDOM	G MO	R DS MAX	C ISS MAX	C RSS MAX	USE	SUPP LIER	EUR SUB	USA SUB	ISS
LDF692	NJD	B60	30V	50MA	150C	.36WH	5MXV			60R	16P		RMS	OBS			0
LDF693	NJD	B60	30V	50MA	150C	.36WH	3MXV			100R	16P		RMS	OBS			0
LS3069	NJD	B10	50V	10MA	125C	0.2WF	10MXV	2/10MA	1/2.5MO		15P		ALG	LED	BFW61	2N3069	0
LS3070	NJD	B10	50V	5MA	125C	0.2WF	5MXV	.5/2.5MA	.75/2.5MO		15P		ALG	LED	BF347	2N3070	0
LS3071	NJD	B10	50V	5MA	125C	0.2WF	2.5MXV	.1/0.6MA	0.5/2.5MO		15P		ALG	LED	BFW13	2N3071	0
LS3458	NJD	B10	50V	25MA	125C	0.2WF	8MXV	3/15MA	2.5/10MO		5P		ALG	LED	BF810	2N3458	0
LS3459	NJD	B10	50V	10MA	125C	0.2WF	4MXV	0.8/4MA	1.5/6MO		5P		ALG	LED	BC264A	2N3459	0
LS3460	NJD	B10	50V	5MA	125C	0.2WF	2MXV	0.2/1MA	0.8/4.5MO		5P		ALG	LED	BF808	2N3460	0
LS3684	NJD	B10	50V	10MA	125C	0.2WF	5MXV	3/7.5MA	2/3MO		4P		ALG	LED	BC264B	2N3684	0
LS3685	NJD	B10	50V	10MA	125C	0.2WF	3.5MXV	1/3.5MA	1.5/2.5MO		4P		ALN	LED	BFW12	2N3685	0
LS3686	NJD	B10	50V	5MA	125C	0.2WF	2MXV	.4/1.2MA	1/2MO		4P		ALN	LED	BF800	2N3686	0
LS3687	NJD	B10	50V	5MA	125C	0.2WF	1.2MXV	.1/.5MA	0.5/1.5MO		4P		ALN	LED	BF800	2N3687	0
LS3819	NJD	B10	25V	25MA	125C	0.2WF	8MXV	2/20MA	2/6.5MO		8P		ALG	LED	BFW61	2N3819	0
LS3821	NJD	B10	50V	10MA	125C	0.2WF	4MXV	1/2.5MA	1.5/4.5MO		6P		ALG	LED	BF347	2N5457	0
LS3822	NJD	B10	50V	25MA	125C	0.2WF	6MXV	0.5/10MA	3/6.5MO		6P		ALG	LED	BFW12	2N3822	0
LS3823	NJD	B10	30V	25MA	125C	0.2WF	8MXV	4/20MA	3.5/6.5MO		6P		FVG	LED	BFW61	2N3823	0
LS3921	NJD	B75	50V	25MA	125C	0.4WF	4MXV	1.5/10MA	1.5MNMO		18P		DUA	LED	BFQ10	2N3921	0
LS3954	NJD	B10	50V	10MA	125C	0.4WF	4.5MXV	0.5/5MA			4P		DUA	LED	BFQ10	2N3954	0
LS3967	NJD	B10	30V	25MA	125C	0.2WF	5MXV	2.5/10MA	2.5MNMO		5P		ALG	LED	BFW12	2N3967	0
LS3968	NJD	B10	30V	10MA	125C	0.2WF	3MXV	1/5MA	2MNMO		5P		ALG	LED	BFW12	2N3968	0
LS3969	NJD	B10	30V	5MA			1.7MXV	0.4/2MA	1.3MNMO		5P		ALG	LED	BFW13	2N3969	0
LS4220	NJD	B10	30V	10MA	125C	0.2WF	4MXV	0.5/3MA	1/4MO		6P		ALG	LED	BF347	2N4220	0
LS4221	NJD	B10	30V	10MA	125C	0.2WF	6MXV	0.2/6MA	2/5MO		6P		ALG	LED	BF347	2N4221	0
LS4222	NJD	B10	30V	25MA			8MXV	0.5/15MA	2.5/6MO		6P		ALG	LED	BFW61	2N4222	0
LS4223	NJD	B10	30V	25MA	125C	0.2WF	8MXV	3/18MA	1.5MNMO		6P		FVG	LED	BFW61	2N4223	0

TYPE NO	CONS TRUC TION	CASE & LEAD	V DS MAX	I D MAX	T J MAX	P TOT MAX	VP OR VT	IDSS OR IDOM	G MO	R DS MAX	C ISS MAX	C RSS MAX	USE	SUPP LIER	EUR SUB	USA SUB	ISS
LS4224	NJD	B10	30V	25MA	125C	0.2WF	8MXV	2/20MA	2/7.5MO		6P		FVG	LED	BFW61	2N4224	0
LS4338	NJD	B10	50V	5MA	125C	0.2WF	1MXV	0.2/.6MA	0.6/1.8MO		6P		ALG	LED	BFW13	2N4338	0
LS4339	NJD	B10	50V	5MA	125C	0.2WF	1.8MXV	.5/1.5MA	0.8/2.4MO		6P		ALG	LED	BFW13	2N4339	0
LS4340	NJD	B10	50V	10MA	125C	0.2WF	3MXV	1/3.6MA	1.3/3MO		6P		ALG	LED	BFW12	2N4340	0
LS4341	NJD	B10	50V	25MA	125C	0.2WF	6MXV	3/9MA	2/4MO		6P		ALG	LED	BFW61	2N4341	0
LS4391	NJD	B10	40V	150MA	125C	0.2WF		50/150MA		30R	14P		RLS	LED	BSV78	2N4391	0
LS4392	NJD	B10	40V	75MA	125C	0.2WF		25/75MA		60R	14P		RLS	LED	BSV79	2N4392	0
LS4393	NJD	B10	40V	50MA	125C	0.2WF		5/30MA		100R	14P		RLS	LED	BSV80	2N4393	0
LS4416	NJD	B10	30V	25MA	125C	0.2WF	6MXV	5/15MA	4/7.5MO		6P		TUG	LED	BF256B	2N4416	0
LS4856	NJD	B10	40V	200MA	125C	0.2WF		50MNMA		25R	18P		RLS	LED	BSV78	2N4856	0
LS4857	NJD	B10	40V	100MA	125C	0.2WF		20/100MA		40R	18P		RLS	LED	BSV79	2N4857	0
LS4858	NJD	B10	40V	80MA				8/80MA		60R	18P		RLS	LED	BSV80	2N4858	0
LS4859	NJD	B10	30V	200MA	125C	0.2WF		50MNMA		25R	18P		RLS	LED	BSV78	2N4859	0
LS4860	NJD	B10	30V	100MA	125C	0.2WF		20/100MA		40R	18P		RLS	LED	BSV79	2N4860	0
LS4861	NJD	B10	30V	80MA	125C	0.2WF		8/80MA		60R	18P		RLS	LED	BSV80	2N4861	0
LS5103	NJD	B10	25V	10MA	125C	0.2WF	4MXV	1/8MA	2/8MO		5P		FVG	LED	BF256A	2N5103	0
LS5104	NJD	B10	25V	10MA	125C	0.2WF	4MXV	2/6MA	3.5/7.5MO		5P		FVG	LED	BF256A	2N5104	0
LS5105	NJD	B10	25V	25MA	125C	0.2WF	4MXV	5/15MA	5/10MA		5P		FVG	LED	BFW61	2N5105	0
LS5245	NJD	B10	30V	25MA	125C	0.2WF	6MXV	5/15MA	4.5/7.5MO		4P5		TUG	LED	BF256B	2N5245	0
LS5246	NJD	B10	30V	10MA	125C	0.2WF	4MXV	1.5/7MA	3/6MO		4P5		FVG	LED	BF256A	2N5246	0
LS5247	NJD	B10	30V	25MA	125C	0.2WF	8MXV	8/24MA	4.5/8MO		4P5		FVG	LED	BF256C	2N5247	0
LS5248	NJD	B10	30V	25MA	125C	0.2WF	8MXV	4/20MA	3.5/6.5MO		6P		FVG	LED	BF256C	2N5248	0
LS5358	NJD	B10	40V	5MA	125C	0.2WF	3MXV	0.5/1MA	1/3MO		8P		ALG	LED	BFW13	2N5358	0
LS5359	NJD	B10	40V	5MA	125C	0.2WF	4MXV	.8/1.6MA	1.2/3.6MO		6P		ALG	LED	BFW13	2N5359	0
LS5360	NJD	B10	40V	10MA	125C	0.2WF	4MXV	1.5/3MA	1.5/4.2MO		6P		ALG	LED	BFW12	2N5360	0

TYPE NO	CONS TRUC TION	CASE & LEAD	V DS MAX	I D MAX	T J MAX	P TOT MAX	VP OR VT	IDSS OR IDOM	G MO	R DS MAX	C ISS MAX	C RSS MAX	USE	SUPP LIER	EUR SUB	USA SUB	ISS
LS5361	NJD	B10	40V	10MA	125C	0.2WF	6MXV	2.5/5MA	1.5/4.5MO		6P		ALG	LED	BC264A	2N5361	0
LS5362	NJD	B10	40V	25MA	125C	0.2WF	7MXV	4/8MA	2/5.5MO		6P		ALG	LED	BC264C	2N5362	0
LS5363	NJD	B10	40V	25MA	125C	0.2WF	8MXV	7/14MA	2.5/6MO		6P		ALG	LED	BC264D	2N5363	0
LS5364	NJD	B10	40V	25MA	125C	0.2WF	8MXV	9/18MA	1.5/2.7MO		6P		ALG	LED	BFW10	2N5364	0
LS5391	NJD	B10	70V	25MA	125C	0.2WF	2MXV	15MXMA	1.5/4.5MO		18P		ALG	LED	BC264D	2N5391	0
LS5392	NJD	B10	70V	10MA	125C	0.2WF	5MXV	3MXMA	2/6MO		18P		ALG	LED	BF347	2N5392	0
LS5393	NJD	B10	70V	10MA	125C	0.2WF	3MXV	4.5MXMA	3/6.5MO		18P		ALG	LED	BFW12	2N5393	0
LS5394	NJD	B10	70V	10MA	125C	0.2WF	4MXV	6MXMA	4/7MO		18P		ALG	LED	BFW56	2N5394	0
LS5395	NJD	B10	70V	25MA	125C	0.2WF	4MXV	8MXMA	4.5/7MO		18P		ALG	LED	BFW56	2N5395	0
LS5396	NJD	B10	70V	25MA	125C	0.2WF	5MXV	10MXMA	4.5/7.5MO		18P		ALG	LED	BFW61	2N5396	0
LS5457	NJD	B10	25V	10MA	125C	0.2WF	6MXV	1/5MA	1/5MO		7P		ALG	LED	BFW12	2N5457	0
LS5458	NJD	B10	25V	25MA	125C	0.2WF	7MXV	2/9MA	1.5/5.5MO		7P		ALG	LED	BC264	2N5458	0
LS5459	NJD	B10	25V	25MA	125C	0.2WF	8MXV	4/16MA	2/6MA		7P		ALG	LED	BC264D	2N5459	0
LS5484	NJD	B10	25V	10MA	125C	0.2WF	3MXV	1/5MA	3/6MA		4P		FVG	LED	BF256A	2N5484	0
LS5485	NJD	B10	25V	25MA	125C	0.2WF	4MXV	4/10MA	3.5/7MO		4P		FVG	LED	BF256B	2N5485	0
LS5486	NJD	B10	25V	25MA	125C	0.2WF	6MXV	8/20MA	4/8MO		4P		FVG	LED	BF256	2N5486	0
LS5556	NJD	B10	30V	10MA	125C	0.2WF	4MXV	.5/2.5MA	1.5/6.5MO		6P		ALG	LED	BF347	2N5556	0
LS5557	NJD	B10	30V	10MA	125C	0.2WF	5MXV	2/5MA	1.5/6.5MO		6P		ALG	LED	BC264A	2N5557	0
LS5558	NJD	B10	30V	25MA	125C	0.2WF	6MXV	4/10MA	1.5/6.5MO		6P		ALN	LED	BFW11	2N5558	0
LS5638	NJD	B10	30V	200MA	125C	0.2WF		50MNMA		30R	10P		RLS	LED	BSV78	2N5638	0
LS5639	NJD	B10	30V	100MA	125C	0.2WF		25MNMA		60R	10P		RLS	LED	BSV80	2N5639	0
LS5640	NJD	B10	30V	25MA	125C	0.2WF		5MNMA		100R	10P		RLS	LED	BSV80	2N5640	0
M100	NMD	B65	20V	12MA	200C	0.3WF	5MXV	4.5MXMA	1/4MO	150R	7P5		RLS	SIU	BFW96	2N3631	0
M100CHP	NMD	B73	20V	20MA	200C		5MXV	1.5/4MA	1/2.2MO	350R	7P5		RLS	SIU			0
M101	NMD	B65	20V	12MA	200C	0.3WF	8MXV	4/12MA	1.5/5MO	100R	7P5		RLS	SIU	BFW96	2N3631	0

TYPE NO	CONS TRUC TION	CASE & LEAD	V DS MAX	I D MAX	T J MAX	P TOT MAX	VP OR VT	IDSS OR IDOM	G MO	R DS MAX	C ISS MAX	C RSS MAX	USE	SUPP LIER	EUR SUB	USA SUB	ISS
M103	PME	B53	30V	50MA	125C	.22WF	2.5/5.5V			130R			RLS	SIU		3N161	0
M103CHP	PME	B73	30V	50MA	125C		2.5/5.5V			100R	8P		RLS	SIU			0
M104	PME	B53	30V	50MA	125C	.22WF	3/6V			1K2			RLS	SIU	BSW95	3N172	0
M106	PME	B38	30V	50MA	125C	.25WF	2/6V	10MNMA	2MNMO	120R	8P	4P	DUA	SIU		3N165	0
M107	PME	B46	30V	50MA	125C	.25WF	2/6V	10MNMA	2MNMO	120R			DUA	SIU		3N189	0
M108	PME	B46	50V	50MA	125C	.25WF	2/8V	10MNMA	2MNMO	120R	8P	4P	DUA	SIU		3N191	0
M113	PME	B53	30V	50MA	125C	.22WF	1/3V			200R	8P		RLS	SIU		3N161	0
M113CHP	PME	B73	30V	50MA	125C		1/3V			200R	8P		RLS	SIU			0
M114	PME	B53	40V	50MA	125C	.22WF	1.5/4.5V	8/35MA	2MNMO	500R	8P		ALG	SIU		3N161	0
M114CHP	PME	B73	40V	50MA	125C		1.5/4.5V	8/35MA	2/4MO	500R	8P		RLS	SIU			0
M116	NME	B53	30V	50MA	125C	.22WF	1/5V	2/20MA		100R	10P	2P5	RLS	INB		3N171	0
M116/D	NME	B73	30V	50MA	125C	.22WF	1/5V	2/20MA		100R	10P	2P5	RLS	INB		3N171	0
M116/W	NME	B74	30V	50MA	125C	.22WF	1/5V	2/20MA		100R	10P	2P5	RLS	INB		3N171	0
M116CHP	NME	B73	30V	50MA	125C		1/5V	2/20MA		200R	8P	2P5	RLS	SIU			0
M117	NME	B56	50V	50MA	125C	.22WF	1/5V	2/20MA		200R	8P		RLS	SIU		2N4351	0
M117CHP	NME	B73	50V	50MA	125C		1/5V	2/20MA		200R	8P	2P5	RLS	SIU			0
M119	PME	B53	75V	50MA	125C	.22WF	2/6V			230R	16P		RLS	SIU		3N161	0
M163	PME	B53	40V	50MA	150C	.38WF	5MXV	30MXMA	2/4MO	250R	2P9		RLS	OBS		3N163	0
M164	PME	B53	30V	50MA	150C	.38WF	5MXV	30MXMA	2/4MO	300R	2P9		RLS	OBS		3N164	0
M511	PME	B53	30V	50MA	125C	0.2WF	6MXV	6MXMA	1MNMO	150R	4P		RLS	OBS	BSW95	2N4352	0
M511A	PME	B53	30V	50MA	125C	0.2WF	6MXV	6MXMA	1MNMO	300R	3P		RLS	OBS	BSW95	3N155A	0
M517	PME	B53	30V	50MA	125C	0.2WF	6MXV			50R	7P		RLS	OBS	BSW95	3N168	0
MEF68	NJD	B75	25V	15MA	125C	.36WF	5MXV	15MXMA	1/6MO		8P		DUA	MCB	BFQ10	2N5562	0
MEF69	NJD	B75	25V	15MA	125C	.36WF	5MXV	15MXMA	1/6MO		8P		DUA	MCB	BFQ10	2N5562	0
MEF70	NJD	B75	25V	15MA	125C	.36WF	5MXV	15MXMA	1/6MO		8P		DUA	MCB	BFQ10	2N5562	0

TYPE NO	CONS TRUC TION	CASE & LEAD	V DS MAX	I D MAX	T J MAX	P TOT MAX	VP OR VT	IDSS OR IDOM	G MO	R DS MAX	C ISS MAX	C RSS MAX	USE	SUPP LIER	EUR SUB	USA SUB	ISS
MEF101	NJD	B50	30V	20MA	125C	.25WF	10MXV	0.2/20MA	0.5MNMO		8P		ALG	OBS	BF244A	2N5457	0
MEF102	NJD	B50	40V	10MA	125C	.25WF	1.5MXV	0.2/1MA	1MNMO		8P		ALG	OBS	BF347	2N5458	0
MEF103	NJD	B50	40V	5MA	125C	.25WF	4MXV	.9/4.5MA	1.5MNMO		8P	3P	ALG	MCB	BF347	2N5457	0
MEF104	NJD	B50	50V				10MXV	4/20MA	2MNMO		8P		ALG	MCB	BC264	2N5458	0
MEF3069	NJD	B50	50V	10MA	125C	.25WF	10MXV	2/10MA	1/2.5MO		15P	4P	ALG	MCB	BC264	2N3069	0
MEF3070	NJD	B50	50V	5MA	125C	.25WF	5MXV	.5/2.5MA	.75/2.5MO		15P	4P	ALG	MCB	BF347	2N3070	0
MEF3071	NJD	B50	50V	5MA	125C	.25WF	5MXV	.1/0.6MA	0.5/2.5MO		15P		ALG	OBS	BFW13	2N3071	0
MEF3458	NJD	B50	50V	20MA	125C	.25WF	8MXV	3/15MA	2.5/10MO		5P		ALG	MCB	BC264	2N3458	0
MEF3459	NJD	B50	50V	5MA	125C	.25WF	4MXV	0.8/4MA	1.5/6MO		5P		ALG	MCB	BF347	2N3459	0
MEF3460	NJD	B50	50V	5MA	125C	.25WF	2MXV	0.2/1MA	0.8/4.5MO		5P		ALG	MCB	BF347	2N3460	0
MEF3684	NJD	B50	50V	10MA	125C	.25WF	5MXV	2.5/8MA	2/3MO		4P		ALN	MCB	BC264B	2N3684	0
MEF3685	NJD	B50	50V	10MA	125C	.25WF	3.5MXV	1/3.5MA	1.5/2.5MO		4P		ALN	MCB	BC264A	2N3685	0
MEF3686	NJD	B50	50V	5MA	125C	.25WF	2MXV	.4/1.2MA	1/2MO		4P		ALN	MCB	BFW13	2N3686	0
MEF3687	NJD	B50	50V	10MA	125C	.25WF	1.2MXV	0.1/.5MA	0.5/1.5MO		4P		ALN	MCB	BFW13	2N3687	0
MEF3819	NJD	B50	25V	20MA	125C	.25WF	8MXV	2/20MA	2/6.5MO		8P	4P	ALG	MCB	BF347	2N3819	0
MEF3821	NJD	B50	50V	5MA	125C	.25WF	4MXV	1/2.5MA	1.5/4.5MO		6P	3P	FVG	MCB	BFW12	2N3821	0
MEF3822	NJD	B50	50V	5MA	125C	.25WF	8MXV	0.5/1MA	3/6.5MO		6P	3P	ALG	MCB	BF347	2N3822	0
MEF3823	NJD	B50	30V	25MA	125C	.25WF	8MXV	4/20MA	3.5/6.5MO		6P	2P	FVG	MCB	BFW61	2N3823	0
MEF3954	NJD	B20	50V	5MA	200C	.25WF	4.5MXV	0.5/5MA	1MNMO		4P		DUA	MCB	BFQ10	2N3954	0
MEF3955	NJD	B20	50V	5MA20	0C	.25WF	4.5MXV	0.5/5MA	1MNMO		4P		DUA	MCB	BFQ11	2N3955	0
MEF3956	NJD	B20	50V	5MA	200C	.25WF	4.5MXV	0.5/5MA	1MNMO		4P		DUA	MCB	BFQ12	2N3956	0
MEF3957	NJD	B20	50V	5MA	200C	.25WF	4.5MXV	0.5/5MA	1MNMO		4P		DUA	MCB	BFQ15	2N3957	0
MEF3958	NJD	B20	50V	5MA	200C	.25WF	4.5MXV	0.5/5MA	1MNMO		4P		DUA	MCB	BFQ15	2N3958	0
MEF3967	NJD	B50	30V	5MA	125C	.25WF	5MXV	2.5/10MA	2.5MNMO		5V	1P3	ALG	MCB	BC264	2N3967	0
MEF3968	NJD	B50	30V	5MA	125C	.25WF	3MXV	1/5MA	2MNMO		5P	1P3	ALG	MCB	BFW12	2N3968	0

TYPE NO	CONS TRUC TION	CASE & LEAD	V DS MAX	I D MAX	T J MAX	P TOT MAX	VP OR VT	IDSS OR IDOM	G MO	R DS MAX	C ISS MAX	C RSS MAX	USE	SUPP LIER	EUR SUB	USA SUB	ISS
MEF3969	NJD	B50	30V	5MA	125C	.25WF	1.7MXV	0.4/2MA	1.3MXMO		5P	1P3	ALG	MCB	BF347	2N3969	0
MEF4220	NJD	B50	30V	3MA	125C	.25WF	4MXV	0.5/3MA	1/4MO		6P	2P	ALG	MCB	BF347	2N4220	0
MEF4221	NJD	B50	30V	10MA	125C	.25WF	6MXV	0.2/6MA	2/5MO		6P	2P	ALG	MCB	BFW12	2N4221	0
MEF4222	NJD	B50	30V	15MA	125	.25WF	8MXV	0.5/15MA	2.5/6MO		6P	2P	ALG	MCB	BFW61	2N4222	0
MEF4223	NJD	B50	30V	18MA	125C	.25WF	8MXV	3/18MA	1.5MNMO		6P	2P	FVG	MCB	BFW61	2N4223	0
MEF4224	NJD	B50	30V	20MA			8MXV	2/20MA	2/7.5MO		6P	2P	FVG	MCB	BF244	2N4224	0
MEF4302	NJD	B50	30V	5MA	125C	.25WF	4MXV	2/5MA	1MNMO		6P	3P	ALG	MCB	BC264A	2N4302	0
MEF4303	NJD	B50	30V	10MA	125C	.25WF	6MXV	4/10MA	2MNMO		6P	3P	ALG	MCB	BFW54	2N4303	0
MEF4304	NJD	B50	30V	15MA	125C	.25WF	10MXV	0.5/15MA	1MNMO		6P	3P	ALG	MCB	BC264	2N4304	0
MEF4338	NJD	B50	50V	5MA	125C	.25WF	1MXV	.2/0.6MA	0.6/1.8MO		6P	3P	ALG	MCB	BF347	2N4338	0
MEF4339	NJD	B50	50V	5MA	125C	.25WF	1.8MXV	.5/1.5MA	0.8/2.4MO		6P	3P	ALG	MCB	BFW13	2N4339	0
MEF4340	NJD	B50	50V	5MA	125C	.25WF	3MXV	1.2/4MA	1.3/3MO		6P	3P	ALG	MCB	BC264A	2N4340	0
MEF4341	NJD	B50	50V	10MA	125C	.25WF	6MXV	3/9MA	2/4MO				ALG	MCB	BC264A	2N4341	0
MEF4391	NJD	B50	40V	150MA	125C	.25WF		50/150MA		30R	14P	3P5	RLS	MCB	BSV78	2N4391	0
MEF4392	NJD	B50	40V	75MA	125C	.25WF		25/75MA		60R	14P	3P5	RLS	MCB	BSV79	2N4392	0
MEF4393	NJD	B50	40V	30MA	125C	.25WF		5/30MA		100R	14P	3P5	RLS	MCB	BSV80	2N4393	0
MEF4416	NJD	B50	30V	15MA	125C	.25WF	8MXV	5/15MA	4/7.5MO		6P	1P	TUG	MCB	BF256B	2N4416	0
MEF4856	NJD	B50	40V	200MA	125C	.25WF		50MNMA		25R	18P	8P	RLS	MCB	BSV78	2N4856	0
MEF4857	NJD	B50	40V	100MA	125C	.25WF		20/100MA		40R	18P	8P	RLS	MCB	BSV79	2N4857	0
MEF4858	NJD	B50	40V	80MA	125C	.25WF		8/80MA		60R	18P	8P	RLS	MCB	BSV80	2N4858	0
MEF4859	NJD	B50	30V	200MA	125C	.25WF		8/80MA		25R	18P	8P	RLS	MCB	BSV78	2N4859	0
MEF4860	NJD	B50	30V	100MA	125C	.25WF		20/100MA		40R	18P	8P	RLS	MCB	BSV79	2N4860	0
MEF4861	NJD	B50	30V	80MA	125C	.25WF		8/80MA		60R	18P	8P	RLS	MCB	BSV80	2N4861	0
MEF5103	NJD	B50	25V	10MA	125C	.25WF	4MXV	1/8MA	2/8MO		5P	1P	FVG	MCB	BFW12	2N5103	0
MEF5104	NJD	B50	25V	10MA	125C	.25WF	4MXV	2/6MA	3.5/7.5MO		5P	1P	FVG	MCB	BFW61	2N5104	0

TYPE NO	CONS TRUC TION	CASE & LEAD	V DS MAX	I D MAX	T J MAX	P TOT MAX	VP OR VT	IDSS OR IDOM	G MO	R DS MAX	C ISS MAX	C RSS MAX	USE	SUPP LIER	EUR SUB	USA SUB	ISS
MEF5105	NJD	B50	25V	15MA	125C	.25WF	4MXV	5/15MA	5/10MO		5P	1P	FVG	MCB	BF811	2N5105	0
MEF5163	NJD	B50	25V	5MA	125C		8MXV	1/4MA	2/9MO		12P	3P	ALG	MCB	BC264A	2N5163	0
MEF5245	NJD	B50	30V	15MA	125C	.25WF	6MXV	5/15MA	4.5/7.5MO		4P5	1P2	TUG	MCB	BF256B	2N5245	0
MEF5246	NJD	B50	30V	7MA	125C	.25WF	4MXV	1.5/7MA	3/6MO		4P5	1P2	FVG	MCB	BF256A	2N5246	0
MEF5247	NJD	B50	30V	24MA	125C	.25WF	8MXV	8/24MA	4.5/8MO		4P5	1P2	FVG	MCB	BF256C	2N5247	0
MEF5248	NJD	B50	30V	20MA	125C	.25WF	8MXV	4/20MA	3.5/6.5MO		6P		FVG	MCB	BFW61	2N5248	0
MEF5358	NJD	B50	40V	1MA	125C	.25WF	3MXV	0.5/1MA	1/3MO		6P	2P	ALG	OBS	BFW13	2N5358	0
MEF5359	NJD	B50	40V	5MA	125C	.25WF	4MXV	.8/1.6MA	1.2/3.6MO		6P	2P	ALG	MCB	BFW13	2N5359	0
MEF5360	NJD	B50	40V	5MA	125C	.25WF	4MXV	1.5/3MA	1.4/4.2MO		6P	2P	ALG	MCB	BFW12	2N5360	0
MEF5361	NJD	B50	40V	5MA	125C	.25WF	6MXV	2.5/5MA	1.5/4.5MO		6P	2P	ALG	MCB	BC264A	2N5361	0
MEF5362	NJD	B50	40V	10MA	125C	.25WF	7MXV	4/8MA	1.8/5.5MO		6P	2P	ALG	MCB	BC264C	2N5362	0
MEF5363	NJD	B50	40V	15MA	125C	.25WF	8MXV	7/14MA	2.5/6MO		6P	2P	ALG	MCB	BC264D	2N5363	0
MEF5364	NJD	B50	40V	20MA	125C	.25WF	8MXV	9/18MA	1.5/2.7MO		6P	2P	ALG	MCB	BC264D	2N5364	0
MEF5391	NJD	B10	70V	2MA	125C	0.3WF	2MXV	1.5MXMA	1.5/4.5MO		18P		ALN	MCB	BF800	2N5392	0
MEF5392	NJD	B10	70V	3MA	125C	0.3WF	5MXV	3MXMA	2/6MO		18P		ALN	MCB	BF808	2N5393	0
MEF5393	NJD	B10	70V	5MA	125C	0.3WF	3MXV	4.5MXMA	3/6.5MO		18P		ALN	MCB	BFW56	2N5394	0
MEF5394	NJD	B10	70V	6MA	125C	0.3WF	4MXV	6MXMA	4/7MO		18P		ALN	MCB	BF810	2N5394	0
MEF5395	NJD	B10	70V	8MA	125C	0.3WF	4MXV	8MXMA	4.5/7MO		18P		ALN	MCB	BF805	2N5395	0
MEF5396	NJD	B10	70V	10MA	125C	0.3WF	5MXV	10MXMA	4.5/7.5MO		18P		ALN	MCB	BF805	2N5396	0
MEF5457	NJD	B50	25V	5MA	125C	.25WF	6MXV	1/5MA	1/5MO		7P	3P	ALG	MCB	BFW12	2N5457	0
MEF5458	NJD	B50	25V	10MA	125C	.25WF	7MXV	2/9MA	1.5/5.5MO		7P	3P	ALG	MCB	BC264	2N5458	0
MEF5459	NJD	B50	25V	20MA	125C	.25WF	8MXV	4/16MA	2/6MO		7P	3P	ALG	MCB	BFW11	2N5459	0
MEF5484	NJD	B50	25V	5MA	125C	.25WF	3MXV	1/5MA	3/6MO		4P	1P2	FVG	MCB	BC264A	2N5484	0
MEF5485	NJD	B50	25V	10MA	125C	.25WF	4MXV	4/10MA	3.5/7MO		4P	1P2	FVG	MCB	BFW61	2N5485	0
MEF5486	NJD	B50	25V	20MA	125C	.25WF	6MXV	8/20MA	4/8MO		4P	1P2	FVG	MCB	BF256B	2N5486	0

TYPE NO	CONS TRUC TION	CASE & LEAD	V DS MAX	I D MAX	T J MAX	P TOT MAX	VP OR VT	IDSS OR IDOM	G MO	R DS MAX	C ISS MAX	C RSS MAX	USE	SUPP LIER	EUR SUB	USA SUB	ISS
MEF5556	NJD	B50	30V	5MA	125C	.25WC	4MXV	.5/2.5MA	1.5/6.5MO		6P	3P	ALG	MCB	BF347	2N5556	0
MEF5557	NJD	B50	30V	5MA	125C	.25WF	5MXV	2/5MA	1.5/6.5MO		6P	3P	ALG	MCB	B264A	2N5557	0
MEF5558	NJD	B50	30V	10MA	125C	.25WF	6MXV	4/10MA	1.5/6.5MO		6P	3P	ALG	MCB	BC264	2N5558	0
MEF5561	NJD	B20	50V	10MA	125C	.25WF	4MXV	1/10MA	1.8MNMO		8P		DUA	MCB	BFQ10	2N5561	0
MEF5562	NJD	B20	50V	10MA	125C	.25WF	4MXV	1/10MA	1.8MNMO		8P		DUA	MCB	BFQ11	2N5562	0
MEF5563	NJD	B20	50V	10MA	125C	.25WF	4MXV	1/10MA	2MNMO		8P		DUA	MCB	BFQ13	2N5563	0
MEM511	PME	B53	30V	50MA	125C	.22WF	3/6V	3MNMA	1MNMO	300R	5P5		RLS	SIU		3N172	0
MEM511C	PME	B53	25V	50MA	100C	.17WF	6MXV		1MNMO	150R	4P		RLS	OBS		3N156A	0
MEM511CCHP	PME	B73	25V	50MA	125C		3/6V	3MNMA	1MNMO	150R	8P		RLS	SIU			0
MEM511CHP	PME	B73	30V	50MA	125C		3/6V	3MNMA	1MNMO	300R	5P5	2P5	RLS	SIU			0
MEM515	PME	B53	30V	700MA	175C	0.5WF	3MXV		12MNMO	15R	50P		RLS	OBS		3N167	0
MEM517	PME	B53	30V	250MA	125C	0.6WF	5MXV	60MNMA	12TPMO	1K	10P		ALG	OBS		3N163	0
MEM517A	PME	B39	30V	250MA	125C	0.6WF	5MXV	60MNMA	12TPMO	1K	10P		ALG	OBS		3N163	0
MEM517B	PME	B21	30V	50MA	175C	0.3WF	5MXV		1.2MNMO				ALG	GIU	BSW95	3N158	0
MEM517C	PME	B16	25V	250MA	100C	.45WF	5MXV		12TPMO	45R	30P		RLS	OBS		3N168	0
MEM520	PME	B53	30V	50MA	125C	0.2WF	6MXV		1MNMO	250R	2P5		RLS	OBS	BSW95	3N163	0
MEM520C	PME	B53	25V	50MA	100C	.17WF	6MXV		1MNMO	150R	4P		RLS	OBS		3N163	0
MEM550	PME	B42	30V	25MA	125C	0.1WF	6MXV	5MXMA	0.5MNMO	180R	1P1		DUA	OBS		3N189	0
MEM550C	PME	B42	25V	25MA	100C	85MWF	6MXV		0.5MNMO	250R	4P		DUA	OBS		3N190	0
MEM551	PME	B42	30V	25MA	125C	0.1WF	6MXV	5MXMA	0.5MNMO	250R	1P1		DUA	OBS		3N190	0
MEM551C	PME	B42	25V	25MA	100C	85MWF	6MXV		0.5MNMO	250R	4P		DUA	OBS		3N190	0
MEM554	NMX	B66	20V	50MA	125C	.15WF	4MXV	30MXMA	8MNMO		7P		ALG	OBS	BSV81	3N128	0
MEM554C	NMX	B66	20V	50MA	125C	.15WF	4MXV	30MXMA	6MNMO		7P		ALG	OBS	BSV81	3N128	0
MEM556	PME	B53	80V	20MA	100C	0.1WF	6MXV		.8/.95MO	700R	0P5		RLS	OBS	BSW95	3N174	0
MEM556C	PME	B53	70V	20MA	100C	0.1WF	6MXV		0.95MNMO		0P7		RLS	OBS	BSW95	3N174	0

TYPE NO	CONS TRUC TION	CASE & LEAD	V DS MAX	I D MAX	T J MAX	P TOT MAX	VP OR VT	IDSS OR IDOM	G MO	R DS MAX	C ISS MAX	C RSS MAX	USE	SUPP LIER	EUR SUB	USA SUB	ISS
MEM557	NMD	B54	20V	30MA	150C	.15WF	4MXV	30MXMA	8MNMO	300R	5P		RLS	OBS	BSV81	3N138	0
MEM557C	NMD	B54	20V	30MA	125C	.22WF	4MXV	30MXMA	6MNMO	200R	5P		RLS	OBS	BSV81	3N138	0
MEM560	PME	B53	35V	50MA	125C	0.3WF	3MXV	15MNMA	2MNMO		9P		ALG	OBS		3N161	0
MEM560C	PME	B53	30V	50MA	100C	0.2WF	3.5MXV		2/3MO	175R	11P		RLS	OBS	BSW95	3N161	0
MEM561	PME	B53	30V	50MA	150C	0.3WF	3MXV		2MNMO	150R	9P		RLS	OBS		3N163	0
MEM561C	PME	B53	25V	50MA	150C	0.3WF	3MXV		2MNMO	150R	9P		RLS	GIU		3N163	0
MEM562	NME	B56	20V	25MA	125C	.22WF	4MXV	3MNMA	0.6MNMO	150R	4P		RLS	OBS		3N171	0
MEM562C	NME	B56	20V	25MA	100C	.18WF	4MXV	5MNMA	1MNMO	150R	5P		RLS	OBS		3N171	0
MEM563	NME	B56	20V		125C	.22WF	4MXV		2MNMO		5P		RLS	OBS		2N4351	0
MEM563C	NME	B56	20V		125C	.22WF	4MXV		2MNMO	100R	6P		RLS	OBS		2N4351	0
MEM564C	NMD	B66	20V	30MA	150C	.22WF	4MXV	30MXMA	8MNMO		8P		ALG		BF352	3N201	0
MEM571C	NMD	B66	20V	30MA	150C	.15WF	4MXV	30MXMA	8MNMO	200R	6P		FVG		BF353	3N202	0
MEM575	PME	B53	25V	300MA	150C	0.3WF	3.5MXV		10MNMO	13R	50P		RMS	GIU		3N167	0
MEM614	NMX	B66	20V	50MA	150C	.22WF	4MXV	20MXMA	6MNMO		8P		ALG	OBS	BF354	3N203	0
MEM616	NMD	B66	25V	50MA	175C	.36WF	4MXV	30MXMA	12MNMO		6P		FVG	GIU	BF351	3N202	0
MEM617	NMD	B66	25V	50MA	175C	.36WF	4MXV	15MXMA	4.4MNMO		6P		FVG	GIU	BF354	3N203	0
MEM618	NMD	B66	25V	50MA	175C	.36WF	4MXV	20MXMA	10MNMO		5P1		FVG	GIU	BF351	3N202	0
MEM620	NMD	B84	25V	30MA	175C	0.5WF	4MXV	18MXMA	14TPMO		7P		FVG	GIU		MPF120	0
MEM621	NMD	B84	25V	30MA	175C	0.5WF	4MXV	30MXMA	16TPMO		6P		FVG	GIU		MPF120	0
MEM622	NMD	B84	25V	30MA	175C	0.5WF	4MXV	20MXMA	14TPMO		7P		FVG	GIU		MPF122	0
MEM640	NMD	B66	25V	50MA	175C	.36WF	4MXV	20MXMA	10/20MO		8P		FVG	GIU		3N204	0
MEM641	NMD	B66	25V	50MA	175C	.36WF	4MXV	18MXMA	10/20MO		6P		FVG	GIU		3N205	0
MEM642	NMD	B66	25V	50MA	175C	.36WF	4MXV	18MXMA	10/20MO		8P		FVG	GIU		3N205	0
MEM655	NMD	B54	20V	50MA	150C	.22WF	4MXV	20MXMA	6MNMO		7P		FVG	OBS		3N128	0
MEM660	NMD	B66	20V	50MA	150C	.15WF	6MXV	10MXMA		30R	7P		RLS	OBS	BF354	3N200	0

TYPE NO	CONS TRUC TION	CASE & LEAD	V DS MAX	I D MAX	T J MAX	P TOT MAX	VP OR VT	IDSS OR IDOM	G MO	R DS MAX	C ISS MAX	C RSS MAX	USE	SUPP LIER	EUR SUB	USA SUB	ISS
MEM667	NMD	B7	20V	20MA	125C	.15WF	4MXV	5MXMA	2TPMO		5P		ALG	GIU		2N3631	0
MEM680	NMD	B66	25V	50MA	175C	.36WF	4MXV	30MXMA	12MNMO		6P		TUG	GIU		3N209	0
MEM681	NMD	B66	25V	50MA	175C	.36WF	4MXV	20MXMA	4.4TPMO		6P		TUG	GIU		3N206	0
MEM682	NMD	B66	25V	50MA	175C	.36WF	4MXV	20MXMA	10MNMO		5P1		TUG	GIU		3N204	0
MEM688	NMD	B66	25V	50MA	175C	.36WF	4MXV	30MXMA	10MNMO		6P		TUG	GIU		3N205	0
MEM689	NMD	B66	25V	50MA	175C	.36WF	4MXV	30MXMA	10MNMO		6P		TUG	GIU		3N205	0
MEM711	NME	B56	25V		150C	.22WF	1.5MXV		2MNMO	100R	6P		RLS	GIU		3N171	0
MEM803	PME	B53	20V	50MA	150C	0.3WF	6.5MXV			200R	9P		RLS	OBS		3N163	0
MEM804	PME	B53	25V	50MA	150C	0.3WF	3MXV		2MNMO	200R	9P		RLS	OBS		3N161	0
MEM805	PME	B53	40V	50MA	150C	0.3WF	5.5MXV		2MNMO	300R	4P5		RLS	OBS		3N161	0
MEM806	PME	B53	40V	50MA	150C	0.3WF	5.5MXV		2MNMO	300R	4P5		RLS	OBS		3N163	0
MEM806A	PME	B53	40V	50MA	150C	0.3WF	5.5MXV		2MNMO	300R	4P5		RLS	GIU		3N163	0
MEM807	PME	B53	40V	50MA	150C	0.3WF	5.5MXV		2MNMO	300R	4P5		RLS	OBS		3N163	0
MEM807A	PME	B53	40V	50MA	150C	0.3WF	5.5MXV		2MNMO	300R	4P5		RLS	GIU		3N172	0
MEM808	PME	B53	40V	50MA	150C	0.3WF	5.5MXV		2MNMO	300R	4P5		RLS	OBS		3N163	0
MEM814	PME	B53	35V	50MA	125C	0.3WF	3MXV		2MNMO	80R	7P		RLS	GIU	BSV81	3N217	0
MEM817	PME	B53	45V	30MA	125C	.22WF	5.5MXV		1MNMO	350R	6P		RLS	GIU		3N164	0
MEM823	PME	B80	45V	3MA	125	.22WF	5.5MXV		1MNMO	350R	6P		RLS	GIU		3N164	0
MEM955	PME	B42	35V	50MA	125C	0.1WF	5MXV		0.5MNMO	100R			RLS	GIU		3N191	0
MEM955A	PME	B42	35V	50MA	125C	0.1WF	5MXV		0.5MNMO	100R			RLS	GIU		3N191	0
MEM955B	PME	B42	35V	50MA	125C	0.1WF	5MXV		0.5MNMO	100R			RLS	GIU		3N191	0
MFE120	NMD	B66	25V	30MA	200C	0.3WF	4MXV	2/18MA	8/18MO		7P	0P05	FVG	MOB	BFS28	3N187	0
MFE121	NMD	B66	25V	30MA	200C	0.3WF	4MXV	5/30MA	10/20MO		6P	0P05	FVG	MOB		3N202	0
MFE122	NMD	B66	25V	30MA	200C	0.3WF	4MXV	2/20MA	8/18MO		7P	0P05	FVG	MOB	BF352	3N140	0
MFE130	NMD	B66	25V	30MA	200C	0.3WF	4MXV	3/30MA	8/20MO		7P	0P05	FVG	MOB	BF351	3N201	0

TYPE NO	CONS TRUC TION	CASE & LEAD	V DS MAX	I D MAX	T J MAX	P TOT MAX	VP OR VT	IDSS OR IDOM	G MO	R DS MAX	C ISS MAX	C RSS MAX	USE	SUPP LIER	EUR SUB	USA SUB	ISS
MFE131	NMD	B66	25V	30MA	200C	0.3WF	4MXV	3/30MA	8/20MO		7P	0P05	FVG	MOB	BF351	3N201	0
MFE132	NMD	B66	25V	30MA	200C	0.3WF	4MXV	3/30MA	8/20MO		7P	0P05	FVG	MOB	BF351	3N201	0
MFE140	NMD	B66	25V	30MA	200C	0.3WF	4MXV	30MXMA	10/20MO		7P		FVG	MOU		3N202	0
MFE590	NME	B66	25V	30MA	200C	0.3WF	0.1MNV		8/20MO		2P5	0P03	TUG	MOB			0
MFE591	NME	B66	25V	30MA	200C	0.3WF	0.1MNV		10/20MO		3P	0P02	TUG	MOB			0
MFE823	PME	B80	25V	30MA	200C	0.3WF	2/6V	3MNMA	1MNMO		6P	1P5	ALG	MOB		3N164	0
MFE824	NMX	B7	20V	30MA	200C	0.3WF	6MXV	1/15MA	1/4MO		4P	0P7	ALG	MOB		2N3797	0
MFE2000	NJD	B15	25V	30MA	175C	0.3WF		4/10MA	2.5/6MO		5P	1P	TUG	MOB	BF256B	2N4416	0
MFE2001	NJD	B15	25V	30MA	175C	0.3WF		8/20MA	4/8MO		5P	1P	TUG	MOB	BF256B	2N4416	0
MFE2004	NJD	B6	30V	150MA	175C	1.8WC	1/6V	8MNMA		80R	16P	5P	RLS	MOB	BSV80	2N4861	0
MFE2005	NJD	B6	30V	150MA	175C	1.8WC	2/8V	15MNMA		50R	16P	5P	RLS	MOB	BSV79	2N4860	0
MFE2006	NJD	B6	30V	150MA	175C	1.8WC	5/10V	30MNMA		30R	16P	5P	RLS	MOB	BSV78	2N4859	0
MFE2007	NJD	B6	25V	250MA	175C	1.8WC	0.5/10V	8MNMA		40R	30P	15P	RLS	MOB	BSV79	2N4860	0
MFE2008	NJD	B6	25V	250MA	175C	1.8WC	1/10V	20MNMA		30R	30P	15P	RLS	MOB	BSV78	2N4859	0
MFE2009	NJD	B6	25V	250MA	175C	1.8WC	3/10V	50MNMA		20R	30P	15P	RLS	MOB		2N4859	0
MFE2010	NJD	B6	25V	300MA	175C	1.8WC	0.5/10V	15MNMA		25R	50P	20P	RLS	MOB		2N4859	0
MFE2011	NJD	B6	25V	300MA	175C	1.8WC	1/10V	40MNMA		15R	50P	20P	RLS	MOB		2N5433	0
MFE2012	NJD	B6	25V	300MA	175C	1.8WC	3/10V	100MNMA		10R	50P	20P	RLS	MOB		2N5432	0
MFE2093	NJD	B14	50V	5MA	175C	0.3WF	2.5MXV	0.1/.7MA		2K5	6P	2P	ALG	MOB	BFW12	2N3821	0
MFE2094	NJD	B14	50V	5MA	175C	0.3WF	4.5MXV	.4/1.4MA		1K6	6P	2P	ALG	MOB	BFW13	2N3821	0
MFE2095	NJD	B14	50V	5MA	175C	0.3WF	5.5MXV	1/3MA		1K3	6P	2P	ALG	MOB	BFW12	2N5457	0
MFE2097	NJD	B23	50V	100MA	175C	1.5WC	7MXV	15/50MA	10/20MO		20P	5P	ALG	MOB			0
MFE2098	NJD	B23	50V	100MA	175C	1.5WC	10MXV	40/100MA	14/25MO		20P	5P	ALG	MOB			0
MFE2133	NJD	B23	30V	100MA	175C	1.5WC	10MXV	25MNMA	12MNMO	60R	20P	5P	RLS	MOB	BSV79	2N4392	0
MFE3001	NMD	B56	20V	20MA	200C	0.3WF	8MXV	0.5/6MA	0.7/3.5MO		5P	1P5	ALG	MOB	BFW96	2N3631	0

TYPE NO	CONS TRUC TION	CASE & LEAD	V DS MAX	I D MAX	T J MAX	P TOT MAX	VP OR VT	IDSS OR IDOM	G MO	R DS MAX	C ISS MAX	C RSS MAX	USE	SUPP LIER	EUR SUB	USA SUB	ISS
MFE3002	NME	B54	15V	30MA	175C	0.3WF	3MXV			100R	5P	1P	RLS	MOB		3N169	0
MFE3003	PME	B54	15V	30MA	175C	0.3WF	4MXV			200R	5P	1P	RLS	MOB		3163	0
MFE3004	NMD	B54	20V	10MA	175C	0.3WF	5MXV	2/10MA	2MNMO		4P5	OP2	TUG	MOB		3N128	0
MFE3005	NMD	B54	20V	10MA	175C	0.3WF	5MXV	2/10MA	2MNMO		4P5	OP2	TUG	MOB		3N128	0
MFE3006	NMD	B66	25V	30MA	175C	0.3WF	3MXV	2/18MA	8/18MO		6P	OP02	FVG	MOB	BF354	3N140	0
MFE3007	NMD	B66	25V	30MA	175C	0.3WF	3MXV	5/20MA	10/18MO		5P5	OP02	FVG	MOB	BF354	3N202	0
MFE3008	NMD	B66	25V	30MA	175C	0.3WF	3MXV	2/20MA	10/18MO		6P	OP02	FVG	MOB	BF354	3N202	0
MFE3020	PME	B38	25V	200MA	175C	0.3WF	2/6V	10/75MA	0.5MNMO	500R	7P	1P5	DUA	MOB		2N4066	0
MFE3021	PME	B38	25V	200MA	175C	0.3WF	2/6V	10/75MA	0.5MNMO	250R	7P	1P5	DUA	MOB		2N4067	0
MFE4007	PJD	B12	40V	20MA	175C	0.2WF	3MXV	0.5/1MA	0.9/2.7MO		7P	2P	ALG	MOB	BF320A	2N5265	0
MFE4008	PJD	B12	40V	20MA	175C	0.2WF	3MXV	.8/1.6MA	1/3MO		7P	2P	ALG	MOB	BF320A	2N5266	0
MFE4009	PJD	B12	40V	20MA	175C	0.2WF	6MXV	1.5/3MA	1.5/3.5MO		7P	2P	ALG	MOB	BF320A	2N5267	0
MFE4010	PJD	B12	40V	20MA	175C	0.2WF	6MXV	2.5/5MA	2/4MO		7P	2P	ALG	MOB	BF320B	2N5268	0
MFE4011	PJD	B12	40V	20MA	175C	0.2WF	8MXV	4/8MA	2.2/4.5MO		7P	2P	ALG	MOB	BF320B	2N5269	0
MFE4012	PJD	B12	40V	20MA	175C	0.2WF	8MXV	7/14MA	2.5/5MO		7P	2P	ALG	MOB	BF320C	2N5270	0
MK10	NJD	B1	20V	20MA	125C	.15WF	8MXV	1/20MA	3MNMO				FVG	MIT	BC264	2N3819	0
MK10D	NJD	B1	20V	20MA	125C	.15WF	8MXV	1/6MA	3MNMO				FVG	MIT	BC264A	2N5484	0
MK10E	NJD	B1	20V	20MA	125C	.15WF	8MXV	5/12MA	3MNMO				FVG	MIT	BC264C	2N5485	0
MK10F	NJD	B1	20V	20MA	125C	.15WF	8MXV	10/20MA	3MNMO				FVG	MIT	BF244	2N5486	0
ML111B	PJD	B62	40V	50MA	125C	0.3WF	6.5MXV			1K6	3P		RLS	OBS	BF320B	2N2386A	0
MM2090	NJD	B57	50V	20MA	175C	0.2WF	2.5MXV	0.2/2MA	0.5/2MO	1K	5P	OP5	RLG	OBS		3N124	0
MM2091	NJD	B57	50V	20MA	175C	0.2WF	4MXV	1/3MA	0.8/2.4MO	750R	5P	OP5	RLG	OBS		3N125	0
MM2092	NJD	B57	50V	20MA	175C	0.2WF	6.5MXV	1.5/5MA	1.8/3.6MO	500R	5P	OP5	RLG	OBS		3N126	0
MM2102	NME	B56	25V	30MA	175C	0.3WF	4MXV	10MXMA	1MNMO	200R	4P5		RLS	OBS		3N171	0
MM2103	PMD	B56	25V	30MA	200C	0.3WF	5MXV	3MNMA	1MNMO	600R	6P5		RLS	OBS			0

TYPE NO	CONS TRUC TION	CASE & LEAD	V DS MAX	I D MAX	T J MAX	P TOT MAX	VP OR VT	IDSS OR IDOM	G MO	R DS MAX	C ISS MAX	C RSS MAX	USE	SUPP LIER	EUR SUB	USA SUB	ISS
MMF1	NJD	B15	30V	20MA	175C	0.3WF	0.2/8V	4/20MA	3.5/6.5MO		6P	2P	MPP	MOB	BFS21A		0
MMF1X2	NJD	B15	30V	20MA	175C	0.3WF	0.2/8V	4/20MA	3.5/6.5MO		6P	2P	MPP	MOB	BFS21A		0
MMF2	NJD	B15	30V	20MA	175C	0.3WF	0.2/8V	4/20MA	3.5/6.5MO		6P	2P	MPP	MOB	BFS21A		0
MMF2X2	NJD	B15	30V	20MA	175C	0.3WF	0.2/8V	4/20MA	3.5/6.5MO		6P	2P	MPP	MOB	BFS21A		0
MMF3	NJD	B15	30V	20MA	175C	0.3WF	0.2/8V	4/20MA	3.5/6.5MO		6P	2P	MPP	MOB	BFS21A		0
MMF3X2	NJD	B15	30V	20MA	175C	0.3WF	0.2/8V	4/20MA	3.5/6.5MO		6P	2P	MPP	MOB	BFS21A		0
MMF4	NJD	B15	30V	20MA	175C	0.3WF	0.2/8V	4/20MA	3.5/6.5MO		6P	2P	MPP	MOB	BFS21A		0
MMF4X2	NJD	B15	30V	20MA	175C	0.3WF	0.2/8V	4/20MA	3.5/6.5MO		6P	2P	MPP	MOB	BFS21A		0
MMF5	NJD	B15	30V	20MA	175C	0.3WF	0.2/8V	4/20MA	3.5/6.5MO		6P	2P	MPP	MOB	BFS21A		0
MMF5X2	NJD	B15	30V	20MA	175C	0.3WF	0.2/8V	4/20MA	3.5/6.5MO		6P	2P	MPP	MOB	BFS21A		0
MMF6	NJD	B15	30V	20MA	175C	0.3WF	0.2/8V	4/20MA	3.5/6.5MO		6P	2P	MPP	MOB	BFS21A		0
MMF6X2	NJD	B15	30V	20MA	175C	0.3WF	0.2/8V	4/20MA	3.5/6.5MO		6P	2P	MPP	MOB	BFS21A		0
MMT3823	NJD	B81	30V	20MA	135C	.22WF	8MXV	5/20MA	3/8MO		4P	1P	FVG	MOB			0
MPF102	NJD	B3	25V	20MA	135C	0.3WF	8MXV	2/20MA	2/7.5MO		7P	3P	RLG	TDY	BFW61	2N5486	0
MPF103	NJD	B3	25V	20MA	135C	0.3WF	0.5/6V	1/5MA	1/5MO		7P	3P	ALG	TDY	BC264A	2N5457	0
MPF104	NJD	B3	25V	10MA	135C	0.3WF	7MXV	2/9MA	1.5/5.5MO		7P	3P	ALG	TDY	BC264C	2N5458	0
MPF105	NJD	B3	25V	16MA	135C	0.3WF	8MXV	4/16MA	2/6MO		7P	3P	ALG	TDY	BC264D	2N5459	0
MPF106	NJD	B3	25V	10MA	125C	0.2WF	0.5/4V	4/10MA	2.5MNMO		5P	1P2	TUG	TDY	BF256A	2N5246	0
MPF107	NJD	B3	25V	20MA	125C	0.2WF	2/6V	8/20MA	4MNMO		5P	1P2	TUG	TDY	BF256C	2N5247	0
MPF108	NJD	B3	25V	24MA	135C	.31WF	0.5/8V	1.5/24MA	1.6MNMO		6P5	2P5	ALN	TDY	BC264	2N3684	0
MPF108/B	NJD	B3	25V	25MA	135C	.31WF	1/7V	7/14MA	2.5/7MO		6P5	2P5	FVG	MOB		2N5245	0
MPF108/G	NJD	B3	25V	25MA	135C	.31WF	1/7V	4/8MA	2.5/7MO		6P5	2P5	FVG	MOB		2N5485	0
MPF108/O	NJD	B3	25V	25MA	135C	.31WF	0.5/5V	1.5/3MA	2/6.5MO		6P5	2P5	FVG	MOB		2N5484	0
MPF108/V	NJD	B3	25V	25MA	135C	.31WF	2/8V	12/24MA	3/7.5MO		6P5	2P5	FVG	MOB		2N5486	0
MPF108/Y	NJD	B3	25V	25MA	135C	.31WF	0.5/5V	2.5/5MA	2/6.5MO		6P5	2P5	FVG	MOB		2N5104	0

TYPE NO	CONS TRUC TION	CASE & LEAD	V DS MAX	I D MAX	T J MAX	P TOT MAX	VP OR VT	IDSS OR IDOM	G MO	R DS MAX	C ISS MAX	C RSS MAX	USE	SUPP LIER	EUR SUB	USA SUB	ISS
MPF109	NJD	B3	25V	24MA	135C	.31WF	0.2/8V	0.5/24MA	0.8/6MO		7P	3P	ALN	TDY	BF244	2N4304	0
MPF109/B	NJD	B3	25V	25MA	135C	.31WF	2/8V	7/14MA	2/6MO		7P	3P	ALG	MOB	BFW61	2N5459	0
MPF109/G	NJD	B3	25V	25MA	135C	.31WF	1/6V	4/8MA	1.5/5MO		7P	3P	ALG	MOB	BFW12	2N4303	0
MPF109/O	NJD	B3	25V	25MA	135C	.31WF	0.4/4V	1.5/3MA	1/4MO		7P	3P	ALG	MOB	BFW12	2N5457	0
MPF109/R	NJD	B3	25V	25MA	135C	.31WF	0.4/4V	.8/1.6MA	1/4MO		7P	3P	ALG	MOB	BFW13	2N3821	0
MPF109/V	NJD	B3	25V	25MA	135C	.31WF	2/8V	12/24MA	2/6MO		7P	3P	ALG	MOB	BF244	2N3822	0
MPF109/W	NJD	B3	25V	25MA	135C	.31WF	0.2/2V	0.5/1MA	0.8/3.2MO		7P	3P	ALG	MOB	BFW13	2N3821	0
MPF109/Y	NJD	B3	25V	25MA	135C	.31WF	1/6V	2.5/5MA	1.5/5MO		7P	3P	ALG	MOB	BFW12	2N5458	0
MPF110	NJD	B3		20MA	125C	.36WF	10MXV	20MXMA	0.5/7.5MO				ALN	SIU	BC264D	2N4869A	0
MPF111	NJD	B3	20V	20MA	125C	0.2WF	0.5/10V	0.5/20MA	0.5MNMO		4P5	1P5	RLG	SIU	BFW61	2N5458	0
MPF112	NJD	B3	25V	25MA	125C	0.2WF	10MXV	1/25MA	1/7.5MO		16P	6P	FVG	SIU	BFW61	2N5458	0
MPF120	NMD	B84	25V	30MA	175C	0.5WC	4MXV	2/18MA	8/18MO		7P	0P5	FVG	MOB			0
MPF121	NMD	B84	25V	30MA	175C	0.5WC	4MXV	5/30MA	10/20MO		6P	0P5	FVG	MOB			0
MPF122	NMD	B84	25V	30MA	175C	0.5WC	4MXV	2/20MA	8/18MO		7P	0P5	FVG	MOB			0
MPF130	NMD	B82	25V	30MA	175C	.35WF	4MXV	3/30MA	8/20MO		7P	0P05	FVG	MOB			0
MPF131	NMD	B82	25V	30MA	175C	.35WF	4MXV	3/30MA	8/20MO		7P	0P05	FVG	MOB			0
MPF132	NMD	B82	25V	30MA	175C	.35WF	4MXV	3/30MA	8/20MO		7P	0P05	FVG	MOB			0
MPF161	PJD	B63	40V	20MA	135C	.31WF	0.2/8V	0.5/14MA	0.8/6MO		7P	2P	ALG	MOB	BF320	2N5462	0
MPF161/B	PJD	B63	40V	20MA	135C	.31WF	2/8V	7/14MA	2/6MO		7P	2P	ALG	MOB	BF320C	2N5270	0
MPF161/G	PJD	B63	40V	20MA	135C	.31WF	2/8V	4/8MA	1.5/5MO		7P	2P	ALG	MOB	BF320B	2N5269	0
MPF161/O	PJD	B63	40V	20MA	135C	.31WF	0.4/4	1.5/3MA	1/4MO		7P	2P	ALG	MOB	BF320A	2N5267	0
MPF161/R	PJD	B63	40V	20MA	135C	.31WF	0.4/41	.8/1.6MA	1/4MO		7P	2P	ALG	MOB	BF320A	2N5266	0
MPF161/W	PJD	B63	40V	20MA	135C	.31WF	0.2/2V	0.5/1MA	0.8/3.2MO		7P	2P	ALG	MOB	BF320A	2N5265	0
MPF161/Y	PJD	B63	40V	20MA	135C	.31WF	0 1/6V	2.5/5MA	1.5/5MO		7P	2P	ALG	MOB	BF320B	2N5268	0
MPF208	NJD	B3	25V	25MA	135C	.31WF	8MXV	0.5MNMA	2/7MO				ALG	OBS	BF347	2N3821	0

TYPE NO	CONS TRUC TION	CASE & LEAD	V DS MAX	I D MAX	T J MAX	P TOT MAX	VP OR VT	IDSS OR IDOM	G MO	R DS MAX	C ISS MAX	C RSS MAX	USE	SUPP LIER	EUR SUB	USA SUB	ISS
MPF209	NJD	B3	25V	5MA	135C	.31WF	8MXV	0.5/24MA	2/7MO				ALG	OBS	BF347	2N3821	0
MPF256	NJD	B3	30V	20MA	150C	.35WF	0.5/7.5V	3/18MA	6MNMO		3P	1P2	TUG	MOB	BF256	2N5247	0
MPF256/G	NJD	B3	30V	20MA	150C	.35WF	0.5/7.5V	6/13MA	6MNMO		3P	1P2	TUG	MOB	BF256B	2N5245	0
MPF256/R	NJD	B3	30V	20MA	150C	.35WF	0.5/7.5V	3/7MA	6MNMO		3P	1P2	TUG	MOB	BF256A	2N5247	0
MPF256/V	NJD	B3	30V	20MA	150C	.35WF	0.5/7.5V	11/18MA	6MNMO		3P	1P2	TUG	MOB	BF256C	2N5247	0
MPF820	NJD	B3	25V	30MA	150C	.62WF	5MXV	10MNMA	20TPMO		15P	3P5	FVG	MOB	BF348	2N5397	0
MPF970	PJD	B3	25V	60MA	150C	.35WF	5/12V	15/60MA		100R	12P	5P	RLS	MOB		2N5115	0
MPF971	PJD	B3	25V	60MA	150C	.35WF	5/12V	2/30MA		250R	12P	5P	RLS	MOB		2N3382	0
MPF4391	NJD	B3	30V	130MA	150C	.62WF	4/10V	60/130MA	20TPMO	30R	10P	3P5	RLS	MOB	BSV78	2N4391	0
MPF4392	NJD	B3	30V	130MA	150C	.62WF	2/5V	25/75MA	17TPMO	60R	10P	3P5	RLS	MOB	BSV79	2N4392	0
MPF4393	NJD	B3	30V	130MA	150C	.62WF	0.5/3V	5/30MA	12TPMO	100R	10P	3P5	RLS	MOB	BSV80	2N4393	0
MT01	PME	B56	40V	150MA		0.2WF	6.2MXV		0.65MNMO	200R			RLS	OBS		3N160	0
MT101B	PME	B56	25V		125C	0.2WF	6.5MXV		.65/.85MO	470R	7P		RLS	OBS	BSW95	2N4352	0
MT102B	PME	B44	25V		125C	0.2WF	6.5MXV		.65/.85MO	470R	7P		DUA	OBS		3N208	0
MTF101	NJD		30V	20MA	125C	0.1WF	10MXV	20MXMA	0.5MNMO	2K	8P		ALG	OBS	BFW61	2N3819	0
MTF102	NJD		40V	20MA	125C	0.1WF	1.5MXV	1MXMA	1MNMO	1K5	8P		ALG	OBS	BFW13	2N3686	0
MTF103	NJD		40V	20MA	125C	0.1WF	4MXV	4.5MXMA	1.5MNMO	1K	8P		ALG	OBS	BFW12	2N5457	0
MTF104	NJD		50V	20MA	125C	0.1WF	10MXV	20MXMA	2MNMO	500R	8P		ALG	OBS	BF244	2N5459	0
NDF9401	NJD	B41	50V	10MA	175C	0.5WF	0.5/4V	0.5/10MA	0.95/2MO		5P	OP02	DUA	NAB	BFQ10	2N5452	0
NDF9402	NJD	B41	50V	10MA	175C	0.5WF	0.5/4V	0.5/10MA	0.95/2MO		5P	OP02	DUA	NAB	BFQ10	2N5452	0
NDF9403	NJD	B41	50V	10MA	175C	0.5WF	0.5/4V	0.5/10MA	0.95/2MO		5P	OP02	DUA	NAB	BFQ10	2N5452	0
NDF9404	NJD	B41	50V	10MA	175C	0.5WF	0.5/4V	0.5/10MA	0.95/2MO		5P	OP02	DUA	NAB	BFQ10	2N5452	0
NDF9405	NJD	B41	50V	10MA	175C	0.5WF	0.5/4V	0.5/10MA	0.95/2MO		5P	OP02	DUA	NAT	BFQ10	2N5452	0
NDF9406	NJD	B51	50V	10MA	175C	.37WF	0.5/4V	0.5/10MA	0.95/2MO		5P	OP02	DUA	NAT	BFQ10	2N5561	0
NDF9407	NJD	B51	50V	10MA	175C	.37WF	0.5/4V	0.5/10MA	0.95/2MO		5P	OP02	DUA	NAT	BFQ10	2N5561	0

TYPE NO	CONS TRUC TION	CASE & LEAD	V DS MAX	I D MAX	T J MAX	P TOT MAX	VP OR VT	IDSS OR IDOM	G MO	R DS MAX	C ISS MAX	C RSS MAX	USE	SUPP LIER	EUR SUB	USA SUB	ISS
NDF9408	NJD	B51	50V	10MA	175C	.37WF	0.5/4V	0.5/10MA	0.95/2MO		5P	OP02	DUA	NAT	BFQ10	2N5562	0
NDF9409	NJD	B51	50V	10MA	175C	.37WF	0.5/4V	0.5/10MA	0.95/2MO		5P	OP02	DUA	NAT	BFQ10	2N5563	0
NDF9410	NJD	B51	50V	10MA	175C	.37WF	0.5/4V	0.5/10MA	0.95/2MO		5P	OP02	DUA	NAB	BFQ10	2N5563	0
NF500	NJD	B15	25V	30MA	150C	0.3WF	8MXV	30MXMA	4.5TPMO	180R	2P5		RLG	NAB	BFW61	2N4224	0
NF501	NJD	B15	15V	30MA	125C	0.2WF	8MXV	30MXMA	4.5TPMO	180R	3P		RLS	NAB	BSV80	2N4393	0
NF506	NJD	B15	25V	15MA	150C	0.3WF	5MXV	15MXMA	2.5/7MO		4P		RLG	NAB	BFW61	2N4223	0
NF510	NJD	B6	30V	30MA	200C	0.3WF	0.5/10V	5MNMA		120R	20P		RLS	TDY		2N4393	0
NF511	NJD	B6	20V	30MA	200C	0.3WF	10MXV	5MNMA		120R	20P		RLS	TDY		2N4393	0
NF520	NJD	B15	30V	10MA	150C	0.3WF	8MXV	10MXMA	0.5MNMO		4P		ALG	NAB	BFW12	2N3684	0
NF521	NJD	B15	30V	10MA	150C	0.3WF	8MXV	2MXMA	0.4MNMO		4P		ALG	NAB	BF800	2N3685	0
NF522	NJD	B15	20V	10MA	125C	0.2WF	8MXV	10MXMA	0.5MNMO		4P		ALG	NAB	BFW56	2N3685	0
NF523	NJD	B15	20V	10MA	125C	0.2WF	8MXV	2MXMA	0.4MNMO		4P		ALG	NAB	BFW13	2N3685	0
NF530	NJD	B6	30V	10MA	150C	0.3WF	8MXV	10MXMA	0.5MNMO		4P		ALG	NAB	BFW61	2N3822	0
NF531	NJD	B6	30V	10MA	150C	0.3WF	8MXV	2MXMA	0.4MNMO		4P		ALG	NAB	BFW12	2N3821	0
NF532	NJD	B6	20V	10MA	125C	0.2WF	8MXV	10MXMA	0.5MNMO		4P		ALG	NAB	BF808	2N3684	0
NF533	NJD	B6	20V	10MA	125C	0.2WF	8MXV	2MXMA	0.4MNMO				ALG	NAB	BFW13	2N3685	0
NF550	NJD	B43	20V	15MA	125C	0.3	4.5MXV	15MXMA	2/7MO		5P		DUA	OBS	BFQ10	2N5545	0
NF580	NJD	B6	25V	200MA	175C	0.3WF	4/12V			5R	25P	13P	RLS	NAB		2N5432	0
NF581	NJD	B6	25V	200MA	175C	0.3WF	4/10V			6R	25P	13P	RLS	NAB		2N5432	0
NF582	NJD	B6	25V	200MA	175C	0.3WF	2/6V			10R	25P	13P	RLS	NAB		2N5433	0
NF583	NJD	B6	25V	200MA	175C	0.3WF	0.5/4V			20R	25P	13P	RLS	NAB		2N5434	0
NF584	NJD	B6	25V	200MA	175C	0.3WF	10MXV			10R	25P	13P	RLS	NAB		2N5433	0
NF585	NJD	B6	25V	200MA	175C	0.3WF	6MXV			20R	25P	13P	RLS	NAB		2N4859	0
NF3819	NJD	B6	25V	20MA	150C	.36WF	8MXV	2/20MA	2/6.5MO		8P	4P	ALG	NAT	BFW61	2N3819	0
NF4302	NJD	B6	30V	15MA	125C	0.3WF	4MXV	5MXMA	0.7MNMO		6P		ALG	NAB	BFW12	2N4302	0

TYPE NO	CONS TRUC TION	CASE & LEAD	V DS MAX	I D MAX	T J MAX	P TOT MAX	VP OR VT	IDSS OR IDOM	G MO	R DS MAX	C ISS MAX	C RSS MAX	USE	SUPP LIER	EUR SUB	USA SUB	ISS
NF4303	NJD	B6	30V	15MA	125C	0.3WF	6MXV	10MXMA	1.4MNMO		6P		ALG	NAB	BFW61	2N4303	0
NF4304	NJD	B6	30V	15MA	125C	0.3WF	10MXV	15MXMA	0.7MNMO		6P		ALG	NAB	BFW61	2N4304	0
NF4445	NJD	B6	25V		200C	0.3WF	2/10V	150MNMA		5R	50P	25P	RMS	TDY		2N5432	0
NF4446	NJD	B6	25V	500MA	200C	0.3WF	2/10V	100MNMA		10R	50P	20P	RMS	TDY		2N5433	0
NF4447	NJD	B6	20V	500MA	200C	0.3WF	2/10V	150MNMA		6R	50P	25P	RMS	TDY		2N5432	0
NF4448	NJD	B6	20V	500MA	200C	0.3WF	2/10V	100MNMA		12R	50P	25P	RMS	TDY		2N5433	0
NF5101	NJD	B15	30V	25MA	150C	.36WF	1MXV	4.5MXMA	8MNMO		12P		ALG	NAT	BF347	2N3687A	0
NF5102	NJD	B15	30V	25MA	150C	.36WF	1.4MXV	13MXMA	11MNMO		12P		ALG	NAT	BF347	2N3686A	0
NF5103	NJD	B15	30V	25MA	150C	.36WF	2.2MXV	24MXMA	12MNMO		12P		ALG	NAT	BF347	2N6451	0
NF5163	NJD	B6	25V	40MA	175C	0.3WF	8MXV	40MXMA	2/9MO	500R	20P		ALG	NAT	BFW61	2N5163	0
NF5457	NJD	B6	25V	16MA	135C	0.3WF	0.5/6V	1/5MA	1/5MO		7P	3P	ALG	NAB	BFW12	2N5457	0
NF5458	NJD	B6	25V	16MA	175C	0.3WF	1/7V	2/9MA	1.5/5.5MO		7P	3P	ALG	NAB	BFW61	2N5458	0
NF5459	NJD	B6	25V	16MA	175C	0.3WF	2/8V	4/16MA	2/6MO		7P	3P	ALG	NAB	BC264C	2N5459	0
NF5484	NJD	B6	25V	20MA	150C	0.3WF	3MXV	5MXMA	3/6MO		5P		FVG	NAT	BFW61	2N5484	0
NF5485	NJD	B6	25V	20MA	150C	0.3WF	4MXV	10MXMA	3.5/7MO		5P		FVG	NAT	BFW61	2N5485	0
NF5486	NJD	B6	25V	20MA	150C	0.3WF	7.5MXV	20MXMA	4/8MO		5P		FVG	NAT	BFW61	2N5486	0
NF5555	NJD	B15	25V	15MA	175C	.36WF	10MXV	15MXMA		150R	5P		RLS	NAB		2N5555	0
NF5638	NJD	B6	30V	250MA	175C	.36WF	12MXV	50MNMA		30R	10P	4P	RLS	NAB	BSV78	2N5638	0
NF5639	NJD	B6	30V	250MA	135C	.36WF	8MXV	25MNMA		60R	10P	4P	RLS	NAB	BSV80	2N5639	0
NF5640	NJD	B6	30V	250MA	135C	.36WF	6MXV	5MNMA		100R	10P	4P	RLS	NAB	BSV80	2N5640	0
NF5653	NJD	B6	30V	250MA	175C	.36WF	12MXV	40MNMA		50R	10P	3P5	RLS	NAB	BSV79	25653	0
NF5654	NJD	B6	30V	250MA	175C	.36WF	8MXV	15MNMA		100R	10P	3P5	RLS	NAB	BSV80	2N5654	0
NF6451	NJD	B6	20V	50MA	175C	.36WF	3.5MXV	20MXMA	15/30MO		25P		ALG	NAT	BF817	2N6451	0
NF6452	NJD	B6	25V	50MA	175C	.36WF	3.5MXV	20MXMA	15/30MO		25P		ALG	NAT	BF817	2N6452	0
NF6453	NJD	B6	20V	50MA	175C	.36WF	5MXV	50MXMA	20/40MO		25P		ALG	NAT	BF817	2N6453	0

TYPE NO	CONS TRUC TION	CASE & LEAD	V DS MAX	I D MAX	T J MAX	P TOT MAX	VP OR VT	IDSS OR IDOM	G MO	R DS MAX	C ISS MAX	C RSS MAX	USE	SUPP LIER	EUR SUB	USA SUB	ISS
NF6454	NJD	B6	25V	50MA	175C	.36WF	5MXV	50MXMA	15/30MO		25P		ALG	NAT	BF817	2N6454	0
NKT80111	NJD	B12	20V	10MA	150C	0.1WF	0.5/6V	0.3/6MA	0.7/3.5MO	450R	3P5		ALG	OBS	BFW12	2N5457	0
NKT80112	NJD	B12	20V	10MA	150C	0.1WF	.65/4.5V	0.45/5MA	0.8/3.2MO	450R	3P5		ALG	OBS	BFW12	2N5457	0
NKT80113	NJD	B12	20V	10MA	150C	0.1WF	.65/4.5V	0.45/5MA	0.8/3.2MO	450R	3P5		ALN	OBS	BF808	2N3685A	0
NKT80211	NJD	B6	10V	10MA	200C	.36WF	0.5TPV	0.15MXMA	0.2/0.7MO	400R	23P		ALG	OBS	BF800	2N3687	0
NKT80212	NJD	B6	10V	10MA	200C	.36WF	0.7TPV	0.1/.3MA	0.4/1.1MO	400R	23P		ALG	OBS	BF800	2N3686	0
NKT80213	NJD	B6	10V	10MA	200C	.36WF	1TPV	0.2/.6MA	0.6/1.5MO	400R	23P		ALG	OBS	BFW13	2N3821	0
NKT80214	NJD	B6	10V	10MA	200C	.36WF	1.5TPV	.5/1.5MA	0.9/2.2MO	400R	23P		ALG	OBS	BF347	2N3821	0
NKT80215	NJD	B6	10V	10MA	200C	.36WF	2.5TPV	1/3MA	1.3/3MO	400R	23P		ALG	OBS	BFW12	2N5457	0
NKT80216	NJD	B6	10V	10MA	200C	.36WF	3.5TPV	2/6MA	1.8/4.2MO	400R	23P		ALG	OBS	BFW12	2N5457	0
NKT80421	NJD	B15	30V	15MA	200C	0.3WF	8MXV	12MXMA	5TPMO		4P		ALG	NEM	BC264D	2N3819	0
NKT80422	NJD	B15	30V	15MA	200C	0.3WF	7MXV	10MXMA	5TPMO		4P		ALG	NEM	BC264D	2N3819	0
NKT80423	NJD	B15	30V	15MA	200C	0.3WF	8MXV	12MXMA	6TPMO		2P		ALG	NEM	BC264D	2N3819	0
NKT80424	NJD	B15	30V	15MA	200C	0.3WF	8MXV	12MXMA	5TPMO		3P		ALG	NEM	BC264D	2N3819	0
NPC108	NJD	B10	25V	25MA	125C	0.2WF	6MXV	25MXMA	4/8MO		5P		ALG	OBS	BF244	2N3819	0
NPC108A	NJD	B10	25V	25MA	125C	0.2WF	5MXV	25MXMA	4/8MO		5P		ALG	OBS	BF244	2N3819	0
NPC211N	NJD	B6	5V	6MA	200C	.36WF	0.5TPV	0.15MXMA	0.2/0.7MO		25P		ALG	OBS	BF800	2N3687	0
NPC212N	NJD	B6	5V	6MA	200C	.36WF	0.7TPV	0.3MXMA	0.4/1.1MO		25P		ALG	OBS	BF800	2N3687	0
NPC213N	NJD	B6	5V	6MA	200C	.36WF	1TPV	0.6MXMA	0.6/1.5MO		25P		ALG	OBS	BF800	2N3687	0
NPC214N	NJD	B6	5V	6MA	200C	.36WF	1.5TPV	1.5MXMA	0.9/2.2MO		25P		ALG	OBS	BFW13	2N3686	0
NPC215N	NJD	B6	5V	6MA	200C	.36WF	2.5TPV	3MXMA	1.3/3MO		25P		ALG	OBS	BF347	2N3685	0
NPC216N	NJD	B6	5V	6MA	200C	.36WF	3.5TPV	6MXMA	1.8/4.2MO		25P		ALG	OBS	BC264A	2N3684	0
OT3	NJD	B21	80V		95C	.09WF	12TPV	0.9MXMA	0.07MNMO		1P		ALG	OBS	BF800	2N3687	0
P102	PJD	B7	30V	5MA	200C	0.3WF	1/4V	.9/4.5MA	1MNMO		17P		PHT	SIU			0
P1003	PJD	B7	50V		200C	0.3WF	3MXV	6MXMA	1/3.5MO		20P		ALG	OBS	BF320A	2N5460	0

TYPE NO	CONS TRUC TION	CASE & LEAD	V DS MAX	I D MAX	T J MAX	P TOT MAX	VP OR VT	IDSS OR IDOM	G MO	R DS MAX	C ISS MAX	C RSS MAX	USE	SUPP LIER	EUR SUB	USA SUB	ISS
P1004	PJD	B7	50V		200C	0.3WF	5MXV	20MXMA	2.5/6MO		20P		ALG	OBS	BF320C	2N5465	0
P1005	PJD	B7	50V		200C	0.3WF	8MXV	25MXMA	3.5/7MO		20P		ALG	OBS	BFT11	2N3382	0
P1027	PJD	B15	30V	50MA	200C	0.3WF	3MXV	6MXMA	0.7/3.5MO		20P		ALG	OBS	BF320B	2N3330	0
P1028	PJD	B15	30V	50MA	200C	0.3WF	5MXV	20MXMA	2.5/8MO		30P		ALG	OBS	BF320C	2N3380	0
P1029	PJD	B15	30V	50MA	200C	0.3WF	8MXV	50MXMA	5MNMO		50P		ALG	OBS	BFT11	2N3386	0
P1069E	PJD	B11	20V	5MA	120C	.27WF	1/4V	1/5MA	3/8MO	600R	40P	5P	RLG	TDY	BF320B	2N3934	0
P1086E	PJD	B10	30V	100MA	120C	.27WF	10MXV	10MNMA		75R	45P	10P	RLS	TDY		2N5115	0
P1087E	PJD	B10	30V	50MA	120C	.27WF	5MXV	5MNMA		150R	45P	10P	RLS	TDY		2N5115	0
P1117E	PJD	B11	25V	10MA	125C	0.3WF	4MXV	4.5/10MA	3.5MNMO	350R	35P	5P	RLG	TDY	BF320C	2N5462	0
P1118E	PJD	B11	25V	5MA	125C	0.3WF	4MXV	2/5MA	2.5MNMO	700R	35P	5P	RLG	TDY	BF320B	2N5268	0
P1119E	PJD	B11	15V		125C	0.3WF	4MXV	0.4/10MA	1.5MNMO	1K	20P	5P	RLG	TDY	BF320	2N5461	0
P236	NJD	B15	40V	8MA	150C	0.3WF	0.7/2V	.4/1.2MA	0.7/2MO		25P	5P	PHT	SIU			0
P237	NJD	B15	40V	8MA	150C	0.3WF	1/3V	1.2/3MA	1/3MO		25P	5P	PHT	SIU			0
P238	NJD	B15	40V	8MA	150C	0.3WF	1.8/5V	7.5MXMA	1.3/4MO		25P	5P	PHT	SIU			0
PF510	PJD	B7	30V	25MA	200C	0.3WF	0.5/10V	5MNMA		200R	15P	4P	RLS	NAB		2N5115	0
PF5101	NJD	B3	30V	25MA	150C	.31WF	1MXV	4.5MXMA	8MNMO				ALG	NAT		2N6451	0
PF5102	NJD	B3	30V	25MA	150C	.31WF	1.4MXV	13MXMA	11MNMO				ALG	NAT		2N6451	0
PF5103	NJD	B3	30V	25MA	150C	.31WF	2.2MXV	24MXMA	12MNMO				ALG	NAT		2N6451	0
PFN3066	NJD	B6	50V	15MA	200C	0.4WF	8MXV	4MXMA	0.4/1MO		2P		PHT	OBS			0
PFN3069	NJD	B6	50V	15MA	200C	0.4WF	8MXV	10MXMA	1/2.5MO		3P		PHT	OBS			0
PFN3458	NJD	B6	50V	15MA	200C	0.4WF	8MXV	15MXMA	2.5/10MO		5P		PHT	OBS			0
PH241N	NJD	B6	25V	30MA	200C	0.3WF	1MXV	3MXMA	2/7MO		13P		PHT	OBS			0
PH242N	NJD	B6	25V	30MA	200C	0.3WF	1.5MXV	6MXMA	3.5/7.5MO		13P		PHT	OBS			0
PH243N	NJD	B6	25V	30MA	200C	0.3WF	2.5MXV	15MXMA	5/10MO		13P		PHT	OBS			0
PH244N	NJD	B6	25V	30MA	200C	0.3WF	3MXV	30MXMA	8/15MO		13P		PHT	OBS			0

TYPE NO	CONS TRUC TION	CASE & LEAD	V DS MAX	I D MAX	T J MAX	P TOT MAX	VP OR VT	IDSS OR IDOM	G MO	R DS MAX	C ISS MAX	C RSS MAX	USE	SUPP LIER	EUR SUB	USA SUB	ISS
PL1091	NJD	B69	50V		175C	.15WF			1.5/4.5MO		6P		FVG	OBS		MMT3823	0
PL1092	NJD	B69	50V		175C	.15WF			3/6.5MO		6P		FVG	OBS		MMT3823	0
PL1093	NJD	B69	30V		175C	.15WF			3.5/6.5MO		6P		FVG	OBS		MMT3823	0
PL1094	NJD	B69	50V		175C	.15WF			6.5MXMO	250R	6P		FVG	OBS		MMT3823	0
PTC151	NJD	B2	25V	20MA	125C	0.2WF		2MNMA	2TPMO				ALG		BF244	2N3819	0
PTC151RT	NJD	B2	25V	20MA	125C	0.2WF		2MNMA	2TPMO				ALG		BF244	2N3819	0
PTC152	NJD	B2	30V	20MA	125C	.35WF		5MNMA	5.5TPMO				ALG		BC264C	2N4303	0
PTC152RT	NJD	B2	30V	20MA	125C	.35WF		5MNMA	5.5TPMO				ALG		BC264C	2N4303	0
PTC161RT	NJD	B3	40V		125C	0.4WF		5MNMA	7TPMO				ALG		BF810	2N6451	0
SC1600	PME	B40	30V		150C		7MXV		0.4MNMO				DUA	OBS		3N165	0
SC1601	PME	B40	30V		150C		7MXV		0.4MNMO				DUA	OBS		3N165	0
SC1611	PME	B53	20V		125C		7MXV		0.8MNMO				ALG	OBS	BSW95	3N157	0
SC1612	PME	B53	20V		125C		7MXV		0.8MNMO				ALG	OBS		3N157	0
SC1613	PME	B53	20V		125C		7MXV		0.8MNMO				ALG	OBS		3N157	0
SC1614	PME	B53			125C		7MXV		0.8MNMO				ALG	OBS		3N157	0
SC1625	PME	B40			150C		6MXV						DUA	OBS		3N165	0
SD5010	PME	B42	30V	25MA	125C	0.1WF	5.5MXV	5TPMA	0.5MNMO	250R	1P1		DUA	OBS		3N189	0
SD5011	PME	B42	30V	25MA	125C	0.1WF	5.5MXV	5TPMA	0.5MNMO	250R	1P1		DUA	OBS		3N190	0
SD5012	PME	B42	80V	25MA	125C	0.1WF	4.5MXV	5TPMA	1.2MNMO	400R			DUA	OBS		3N208	0
SD5013	PME	B42	80V	25MA	125C	0.1WF	4.5MXV	5TPMA	1.2MNMO	400R			DUA	OBS		3N208	0
SD5014	PME	B42	120V	25MA	125C	0.1WF	4.5MXV	5TPMA	1.2MNMO				DUA	OBS		3N208	0
SD5015	PME	B42	120V	25MA	125C	0.1WF	4.5MXV	5TPMA	1.2MNMO				DUA	OBS		3N208	0
SD5050	NME	B42	25V	25MA	125C	.11WF	5.5MXV		0.5MNMO	250R	1P5		DUA	OBS			0
SD5051	NME	B42	25V	25MA	125C	.11WF	5.5MXV		0.5MNMO	250R	1P5		DUA	OBS			0
SDF500	NJD	B41	50V	5MA	200C	.25WF	1/4.5V	0.5/5MA	3MXMO		8P	1P8	DUA	SOT	BFQ10	2N5452	0

TYPE NO	CONS TRUC TION	CASE & LEAD	V DS MAX	I D MAX	T J MAX	P TOT MAX	VP OR VT	IDSS OR IDOM	G MO	R DS MAX	C ISS MAX	C RSS MAX	USE	SUPP LIER	EUR SUB	USA SUB	ISS
SDF501	NJD	B41	50V	5MA	200C	.25WF	1/4.5V	0.5/5MA	3MXMO		8P	1P8	DUA	SOT	BFQ10	2N5452	0
SDF502	NJD	B41	50V	5MA	200C	.25WF	1/4.5V	0.5/5MA	3MXMO		8P	1P8	DUA	SOT	BFQ10	2N5452	0
SDF503	NJD	B41	50V	5MA	200C	.25WF	1/4.5V	0.5/5MA	3MXMO		8P	1P8	DUA	SOT	BFQ10	2N5452	0
SDF504	NJD	B41	50V	5MA	200C	.25WF	1/4.5V	0.5/5MA	3MXMO		8P	1P8	DUA	SOT	BFQ10	2N5452	0
SDF505	NJD	B41	50V	5MA	200C	.25WF	1/4.5V	0.5/5MA	3MXMO		8P	1P8	DUA	SOT	BFQ10	2N5452	0
SDF506	NJD	B41	50V	5MA	200C	.25WF	1/4.5V	0.5/5MA	3MXMO		8P	1P8	DUA	SOT	BFQ10	2N5452	0
SDF507	NJD	B41	50V	5MA	200C	.25WF	1/4.5V	0.5/5MA	3MXMO		8P	1P8	DUA	SOT	BFQ10	2N5452	0
SDF508	NJD	B41	50V	5MA	200C	.25WF	1/4.5V	0.5/5MA	3MXMO		8P	1P8	DUA	SOT	BFQ10	2N5452	0
SDF509	NJD	B41	50V	5MA	200C	.25WF	1/4.5V	0.5/5MA	3MXMO		8P	1P8	DUA	SOT	BFQ10	2N5452	0
SDF510	NJD	B41	50V	5MA	200C	.25WF	1/4.5V	0.5/5MA	3MXMO		8P	1P8	DUA	SOT	BFQ10	2N5452	0
SDF511	NJD	B41	50V	5MA	200C	.25WF	1/4.5V	0.5/5MA	3MXMO		8P	1P8	DUA	SOT	BFQ10	2N5452	0
SDF512	NJD	B41	50V	5MA	200C	.25WF	1/4.5V	0.5/5MA	3MXMO		8P	1P8	DUA	SOT	BFQ10	2N5452	0
SDF513	NJD	B41	50V	5MA	200C	.25WF	1/4.5V	0.5/5MA	3MXMO		8P	1P8	DUA	SOT	BFQ10	2N5452	0
SDF514	NJD	B41	50V	5MA	200C	.25WF	1/4.5V	0.5/5MA	3MXMO		8P	1P8	DUA	SOT	BFQ10	2N5452	0
SES3819	NJD	B1	25V	20MA	125C	0.2WF	7.5MXV	20MXMA	2/6.5MO		8P		ALG	OBS	BF244	2N3819	0
SFF121	PME	B54	20V	20MA	125C	0.2WF	3.5MXV		0.7MNMO	1K5		OP5	ALN	THS	BSW95	2N4352	0
SFF122	PME	B54	20V	20MA	125C	0.2WF	3.5MXV		0.7MNMO	1K5		OP5	RLS	THS	BSW95	3N161	0
SFF123	PME	B54	20V	20MA	125C	0.2WF	5MXV		0.7MNMO	1K5		OP6	RLS	THS	BSW95	3N161	0
SFF1104	PME	B54	25V	50MA	125C	0.2WF	5MXV		1MNMO	300R	6P		RLS	THS		3N164	0
SFT601	NJD	B56	40V	10MA	175C	.15WF	5MXV	.25MXMA	0.35/1MO				ALG	OBS	BF800	2N3687A	0
SFT602	NJD	B56	40V	10MA	175C	.15WF	10MXV	1.5MXMA	0.35/1MO				ALG	OBS	BF800	2N4867A	0
SFT603	NJD	B56	40V	10MA	175C	.15WF	20MXV	5MXMA	0.35/1MO				ALG	OBS	BF808	2N4868A	0
SFT604	NJD	B56	30V	10MA	175C	.15WF			0.35/1MO	200R			ALG	OBS	BFW13	2N3966	0
SI211N	NJD	B6	5V	6MA	200C	.36WF	0.5TPV	0.15MXMA	0.2/0.7MO		40P		ALG	OBS	BF800	2N3687	0
SI211NA	NJD	B6	5V	6MA	150C	.36WF	0.5TPV	0.15MXMA	0.2/0.7MO		8P		ALG	OBS	BF800	2N3687	0

TYPE NO	CONS TRUC TION	CASE & LEAD	V DS MAX	I D MAX	T J MAX	P TOT MAX	VP OR VT	IDSS OR IDOM	G MO	R DS MAX	C ISS MAX	C RSS MAX	USE	SUPP LIER	EUR SUB	USA SUB	ISS
SI212N	NJD	B6	5V	6MA	200C	.36WF	0.7TPV	0.3MXMA	0.4/1.1MO		40P		ALG	OBS	BF800	2N3687	0
SI212NA	NJD	B6	5V	6MA	150C	.36WF	0.7TPV	0.3MXMA	0.4/1.1MO		8P		ALG	OBS	BF800	2N3687	0
SI213N	NJD	B6	5V	6MA	200C	.36WF	1TPV	0.6MXMA	0.6/1.5MO		40P		ALG	OBS	BF800	2N3687	0
SI213NA	NJD	B6	5V	6MA	150C	.36WF	1TPV	0.6MXMA	0.6/1.5MO		8P		ALG	OBS	BF800	2N3687	0
SI214N	NJD	B6	5V	6MA	200C	.36WF	1.5TPV	1.5MXMA	0.9/2.2MO		40P		ALG	OBS	BF800	2N3687	0
SI214NA	NJD	B6	5V	6MA	150C	.36WF	1.5TPV	1.5MXMA	0.9/2.2MO		8P		ALG	OBS	BF800	2N3687	0
SI215N	NJD	B6	5V	6MA	200C	.36WF	2.5TPV	3MXMA	1.3/3MO		40P		ALG	OBS	BC264A	2N3685	0
SI215NA	NJD	B6	5V	6MA	150C	.36WF	2.5TPV	3MXMA	1.3/3MO		8P		ALG	MOB	BC264A	2N3685	0
SI216N	NJD	B6	5V	6MA	200C	.36WF	3.5TPV	6MXMA	1.8/4.2MO		40P		ALG	OBS	BC264B	2N3684	0
SI216NA	NJD	B6	5V	6MA	150C	.36WF	3.5TPV	6MXMA	1.8/4.2MO		8P		ALG	OBS	BC264B	2N3684	0
SI221N	NJD	B57	8V	6MA	200C	0.3WF	0.3MXV	0.15MXMA	0.4/1.2MO		30P		ALG	OBS	BF800	2N3687	0
SI222N	NJD	B57	8V	6MA	200C	0.3WF	0.5MXV	0.3MXMA	0.7/1.8MO		30P		ALG	OBS	BF800	2N3687	0
SI223N	NJD	B57	8V	6MA	200C	0.3WF	0.7MXV	0.6MXMA	1/2.4MO		30P		ALG	OBS	BF800	2N3687	0
SI224N	NJD	B57	8V	6MA	200C	0.3WF	1MXV	1.5MXMA	1.6/3.6MO		30P		ALG	OBS	BF800	2N3687	0
SI225N	NJD	B57	8V	6MA	200C	0.3WF	1.5MXV	3MXMA	2.2/5MO		30P		ALG	OBS	BC264A	2N3685	0
SI226N	NJD	B57	8V	6MA	200C	0.3WF	2.5MXV	6MXMA	3/7MO		30P		ALG	OBS	BC264B	2N3684	0
SI231N	NJD	B6	15V	6MA	200C	0.2WF	0.7MXV	0.15MXMA	0.15/.5MO		6P		ALG	OBS	BF800	2N3687	0
SI232N	NJD	B6	15V	6MA	200C	0.2WF	1MXV	0.3MXMA	0.28/.8MO		6P		ALG	OBS	BF800	2N3687	0
SI233N	NJD	B6	15V	6MA	200C	0.2WF	1.4MXV	0.6MXMA	0.4/1MO		6P		ALG	OBS	BF800	2N3687	0
SI234N	NJD	B6	15V	6MA	200C	0.2WF	2MXV	1.5MXMA	.65/1.5MO		6P		ALG	OBS	BFW13	2N3686	0
SI235N	NJD	B6	15V	6MA	200C	0.2WF	3.5MXV	3MXMA	0.9/2MO		6P		ALG	OBS	BF347	2N3685	0
SI236N	NJD	B6	15V	6MA	200C	0.2WF	5MXV	6MXMA	1.3/3MO		6P		ALG	OBS	BFW12	2N3684	0
SI241N	NJD	B6	25V	75MA	200C	0.3WF	1MXV	3MXMA	2/7MO		18P		ALG	OBS	BFW12	2N3821	0
SI242N	NJD	B6	25V	75MA	200C	0.3WF	1.5MXV	6MXMA	3.5/7.5MO		18P		ALG	OBS	BFW56	2N3822	0
SI243N	NJD	B6	25V	75MA	200C	0.3WF	2.5MXV	15MXMA	5/10MO		18P		ALG	OBS	BF810	2N6451	0

TYPE NO	CONS TRUC TION	CASE & LEAD	V DS MAX	I D MAX	T J MAX	P TOT MAX	VP OR VT	IDSS OR IDOM	G MO	R DS MAX	C ISS MAX	C RSS MAX	USE	SUPP LIER	EUR SUB	USA SUB	ISS
SI244N	NJD	B6	25V	75MA	200C	0.3WF	3MXV	30MXMA	8MNMO	100R	13P		RLS	OBS	BSV80	2N4393	0
SI245N	NJD	B6	25V	75MA	200C	0.3WF	5MXV	35MXMA	8MNMO	80R	13P		RLS	OBS	BSV80	2N4093	0
SI246N	NJD	B6	25V	75MA	200C	0.3WF	10MXV	75MXMA	8MNMO	60R	13P		RLS	OBS	BSV80	2N4861	0
SK3050	NMD	B66		50MA	175C	.33WF		15MXMA	12TPMO				FVG		BF354	3N203	0
SK3050RT	NMD	B66		50MA	175C	.33WF		15MXMA	12TPMO				FVG		BF354	3N203	0
SK3065RT	NMD	B66		50MA	175C	.33WF		15MXMA	13TPMO				FVG	RCU	BF354	3N203	0
SU2000	NJD	B6	50V				4MXV		7.5MXMO				ALG	OBS	BC264D	2N3822	0
SU2020	NJD	B51	50V		175C								DUA	OBS	BFQ10	2N5452	0
SU2021	NJD	B51	50V		175C								DUA	OBS	BFQ10	2N5452	0
SU2022	NJD	B51	50V		175C								DUA	OBS	BFQ10	2N5452	0
SU2023	NJD	B51	50V		175C								DUA	OBS	BFQ10	2N5452	0
SU2024	NJD	B51	50V		175C								DUA	OBS	BFQ10	2N5452	0
SU2025	NJD	B51	50V		175C								DUA	OBS	BFQ10	2N5452	0
SU2026	NJD	B51	50V		175C								DUA	OBS	BFQ10	2N5452	0
SU2027	NJD	B51	50V		175C								DUA	OBS	BFQ10	2N5452	0
SU2028	NJD	B51	50V		175C								DUA	OBS	BFQ10	2N5452	0
SU2029	NJD	B51	50V		175C								DUA	OBS	BFQ10	2N5452	0
SU2030	NJD	B51	50V		175C	0.3WF			0.3MNMO				DUA	INB		2N4082	0
SU2031	NJD	B51	50V		175C	0.3WF			0.4MNMO				DUA	INB	BFQ10	2N4082	0
SU2032	NJD	B51	50V		175C	0.3WF			1.5MNMO				DUA	OBS	BFQ10	2N4082	0
SU2033	NJD	B51	50V		175C	0.3WF			2.5MNMO				DUA	INB	BFQ10	2N5561	0
SU2034	NJD	B51	50V		175C	0.3WF			1.5MNMO				DUA	OBS	BFQ10	2N5561	0
SU2035	NJD	B51	50V		175C	0.3WF			2.5MNMO				DUA	INB	BFQ10	2N5561	0
SU2074	NJD	B51	50V	2MA	200C	.25WF	3MXV	1.3MXMA	0.3MNMO		7P		DUA	OBS	BFQ10	2N4082	0
SU2075	NJD	B51	50V	2MA	200C	.25WF	3MXV	1.3MXMA	0.3MNMO		7P		DUA	OBS	BFQ10	2N4082	0

TYPE NO	CONS TRUC TION	CASE & LEAD	V DS MAX	I D MAX	T J MAX	P TOT MAX	VP OR VT	IDSS OR IDOM	G MO	R DS MAX	C ISS MAX	C RSS MAX	USE	SUPP LIER	EUR SUB	USA SUB	ISS
SU2075X2	NJD	B51	50V	2MA	200C	.25WF	3MXV	1.3MXMA	0.3MNMO		7P		DUA	OBS	BFQ10	2N4082	0
SU2076	NJD	B51	50V	10MA	200C	0.3WF	3MXV	10MXMA	1.5MNMO		18P		DUA	OBS	BFQ10	2N5561	0
SU2076X2	NJD	B51	50V	10MA	200C	0.3WF	3MXV	10MXMA	1.5MNMO		18P		DUA	OBS	BFQ10	2N5561	0
SU2077	NJD	B51	50V	10MA	200C	0.3WF	3MXV	10MXMA	1.5MNMO		18P		DUA	OBS	BFQ10	2N5561	0
SU2077X2	NJD	B51	50V	10MA	200C	0.3WF	3MXV	10MXMA	1.5MNMO		18P		DUA	OBS	BFQ10	2N5561	0
SU2078	NJD	B51	50V	2MA	180C	.24WF	4MXV	0.25/2MA	0.3MNMO		7P	2P2	DUA	TDY		2N5902	0
SU2078X2	NJD	B51	50V	2MA	180C	.24WF	4MXV	0.25/2MA	0.3MNMO		7P	2P2	DUA	TDY		2N5902	0
SU2079	NJD	B51	50V	2MA	180C	.24WF	4MXV	0.25/2MA	0.3MNMO		7P	2P2	DUA	TDY		2N4082	0
SU2079X2	NJD	B51	50V	2MA	180C	.24WF	4MXV	0.25/2MA	0.3MNMO		7P	2P2	DUA	TDY		2N4082	0
SU2080	NJD	B51	50V	10MA	180C	.24WF	4MXV	1/10MA	1.5MNMO		18P	6P	DUA	TDY		2N5545	0
SU2080X2	NJD	B51	50V	10MA	180C	.24WF	4MXV	1/10MA	1.5MNMO		18P	6P	DUA	TDY		2N5545	0
SU2081	NJD	B51	50V	10MA	180C	.24WF	4MXV	1/10MA	1.5MNMO		18P	6P	DUA	TDY		2N5545	0
SU2081X2	NJD	B51	50V	10MA	180C	.24WF	4MXV	1/10MA	1.5MNMO		18P	6P	DUA	TDY		2N5545	0
SU2098	NJD	B51	30V	8MA	180C	.24WF	4MXV	1/8MA	1MNMO		7P	2P2	DUA	TDY		2N5545	0
SU2098A	NJD	B51	50V	8MA	180C	.24WF	0.5/4V	1/8MA	1.5/4.5MO		6P	2P	DUA	TDY		2N5545	0
SU2098AX2	NJD	B51	50V	8MA	180C	.24WF	0.5/4V	1/8MA	1.5/4.5MO		6P	2P	DUA	TDY		2N5545	0
SU2098B	NJD	B51	50V	8MA	180C	.24WF	0.5/4V	1/8MA	1.5/4.5MO		6P	2P	DUA	TDY		2N5515	0
SU2098BX2	NJD	B51	50V	8MA	180C	.24WF	0.5/4V	1/8MA	1.5/4.5MO		6P	2P	DUA	TDY		2N5515	0
SU2098X2	NJD	B51	30V	8MA	180C	.24WF	4MXV	1/8MA	1MNMO		7P	2P2	DUA	TDY		2N5545	0
SU2099	NJD	B51	30V	8MA	180C	.24WF	4MXV	1/8MA	1MNMO		7P	2P2	DUA	TDY		2N5545	0
SU2099A	NJD	B51	50V	8MA	180C	.24WF	0.5/4V	1/8MA	1.5/4.5MO		6P	2P	DUA	TDY		2N5545	0
SU2099AX2	NJD	B51	50V	8MA	180C	.24WF	0.5/4V	1/8MA	1.5/4.5MO		6P	2P	DUA	TDY		2N5545	0
SU2099X2	NJD	B51	30V	8MA	180C	.24WF	4MXV	1/8MA	1MNMO		7P	2P2	DUA	TDY		2N5545	0
SU2365	NJD	B51	30V	10MA	180C	.24WF	3.5MXV	0.5/10MA	1/2MO		16P	4P	DUA	TDY		2N5545	0
SU2365A	NJD	B51	30V	10MA	150C	0.5WF	3.5MXV	0.5/10MA	1/2MO		16P	4P	DUA	INB	BFQ10	2N5515	0

TYPE NO	CONS TRUC TION	CASE & LEAD	V DS MAX	I D MAX	T J MAX	P TOT MAX	VP OR VT	IDSS OR IDOM	G MO	R DS MAX	C ISS MAX	C RSS MAX	USE	SUPP LIER	EUR SUB	USA SUB	ISS
SU2366	NJD	B51	30V	10MA	150C	0.5WF	3.5MXV	0.5/10MA	1/2MO		16P	4P	DUA	INB	BFQ10	2N5515	0
SU2366A	NJD	B51	30V	10MA	150C	0.5WF	3.5MXV	0.5/10MA	1/2MO		16P	4P	DUA	INB	BFQ10	2N5515	0
SU2367	NJD	B51	30V		150C	0.5WF	3.5MXV	0.5/10MA	1/2MO		16P	4P	DUA	INB	BFQ10	2N5515	0
SU2367A	NJD	B51	30V	10MA	150C	0.5WF	3.5MXV	0.5/10MA	1/2MO		16P	4P	DUA	INB	BFQ10	2N5515	0
SU2368	NJD	B51	30V	10MA	150C	0.5WF	3.5MXV	0.5/10MA	1/2MO		16P	4P	DUA	INB	BFQ10	2N5515	0
SU2368A	NJD	B51	30V	10MA	150C	0.5WF	3.5MXV	0.5/10MA	1/2MO		16P	4P	DUA	INB	BFQ10	2N5515	0
SU2369	NJD	B51	30V	10MA	150C	0.5WF	3.5MXV	0.5/10MA	1/2MO		16P	4P	DUA	INB	BFQ10	2N5515	0
SU2369A	NJD	B51	30V	10MA	150C	0.5WF	3.5MXV	0.5/10MA	1/2MO		16P	4P	DUA	INB	BFQ10	2N5515	0
SU2410	NJD	B51	40V	1MA	180C	.24WF	3.5MXV	0.1/1MA	.25/1.5MO		3P	1P5	DUA	TDY		2N5902	0
SU2411	NJD	B51	40V	1MA	180C	.24WF	3.5MXV	0.1/1MA	.25/1.5MO		3P	1P5	DUA	TDY		2N4082	0
SU2412	NJD	B51	40V	1MA	180C	.24WF	3.5MXV	0.1/1MA	.25/1.5MO		3P	1P5	DUA	TDY		2N5902	0
T1317	PME	B53	40V	20MA	135C	0.3WF	6MXV		1.4MNMO		3P		ALG	OBS	BSW95	3N164	0
TD5452	NJD	B51	50V	5MA	180C	.24WF	1/4.5V	0.5/5MA	1/3MO		6P	2P	DUA	TDY		2N5916	0
TD5453	NJD	B51	50V	5MA	180C	.24WF	1/4.5V	0.5/5MA	1/3MO		6P	2P	DUA	TDY		2N5196	0
TD5454	NJD	B51	50V	5MA	180C	.24WF	1/4.5V	0.5/5MA	1/3MO		6P	2P	DUA	TDY		2N5196	0
TD5902	NJD	B51	40V	1MA	180C	.24WF	2.5MXV	0.1/.5MA	0.4/1.5MO		3P	1P5	DUA	TDY		2N5902	0
TD5902	NJD	B51	40V	1MA	180C	.24WF	2.5MXV	0.1/.5MA	0.4/1.5MO		3P	1P5	DUA	TDY		2N5902	0
TD5902A	NJD	B51	40V	2MA	180C	.24WF	4.5MXV	0.25/2MA	.025/1MO		3P	1P5	DUA	TDY		2N5902A	0
TD5903	NJD	B51	40V		180C	.24WF	2.5MXV	0.1/.5MA	0.4/1.5MO		3P	1P5	DUA	TDY		2N5903	0
TD5903A	NJD	B51	40V	2MA	180C	.24WF	4.5MXV	0.25/2MA	.025/1MO		3P	1P5	DUA	TDY		2N5903A	0
TD5904	NJD	B51	40V	1MA	180C	.24WF	2.5MXV	0.1/.5MA	0.4/1.5MO		3P	1P5	DUA	TDY		2N5904	0
TD5904A	NJD	B51	40V	2MA	180C	.24WF	4.5MXV	0.25/2MA	.025/1MO		3P	1P5	DUA	TDY		2N5904A	0
TD5905	NJD	B51	40V	1MA	180C	.24WF	2.5MXV	0.1/.5MA	0.4/1.5MO		3P	1P5	DUA	TDY		2N5905	0
TD5905A	NJD	B51	40V	2MA	180C	.24WF	4.5MXV	0.25/2MA	.025/1MO		3P	1P5	DUA	TDY		2N5905A	0
TD5906	NJD	B51	40V	1MA	180C	.24WF	2.5MXV	0.1/.5MA	0.4/1.5MO		3P	1P5	DUA	TDY		2N5906	0

TYPE NO	CONS TRUC TION	CASE & LEAD	V DS MAX	I D MAX	T J MAX	P TOT MAX	VP OR VT	IDSS OR IDOM	G MO	R DS MAX	C ISS MAX	C RSS MAX	USE	SUPP LIER	EUR SUB	USA SUB	ISS
TD5906A	NJD	B51	40V	2MA	180C	.24WF	4.5MXV	0.25/2MA	.025/1MO		3P	1P5	DUA	TDY		2N5906A	0
TD5907	NJD	B51	40V	1MA	180C	.25WF	2.5MXV	0.1/.5MA	0.4/1.5MO		3P	1P5	DUA	TDY		2N5907	0
TD5907A	NJD	B51	40V	2MA	180C	.25WF	4.5MXV	0.25/2MA	.025/1MO		3P	1P5	DUA	TDY		2N5907A	0
TD5908	NJD	B51	40V	1MA	180C	.25WF	2.5MXV	0.1/.5MA	0.4/1.5MO		3P	1P5	DUA	TDY		2N5908	0
TD5908A	NJD	B51	40V	2MA	180C	.25WF	4.5MXV	0.25/2MA	.025/1MO		3P	1P5	DUA	TDY		2N5908A	0
TD5909	NJD	B51	40V	1MA	180C	.25WF	2.5MXV	0.1/.5MA	0.4/1.5MO		3P	1P5	DUA	TDY		2N5909	0
TD5909A	NJD	B51	40V	2MA	180C	.25WF	4.5MXV	0.25/2MA	.025/1MO		3P	1P5	DUA	TDY		2N5909A	0
TD5911	NJD	B51	25V	40MA	180C	.25WF	1/5V	7/40MA	5/10MA		5P	1P2	DUA	TDY		2N5911	0
TD5911A	NJD	B51	30V	40MA	180C	.25WF	1/6V	7/40MA	3.5MNMO		4P	1P2	DUA	TDY		2N5911	0
TD5912	NJD	B51	25V	40MA	180C	.25WF	1/5V	7/40MA	5/10MO		5P	1P2	DUA			2N5912	0
TD5912A	NJD	B51	30V	40MA	180C	.25WF	1/6V	6/40MA	3.5MNMO		4P	1P2	DUA	TDY		2N5912	0
TIS05	PJD	B12	25V			0.3WF		10/45MA	6/12MO	150R	5P		RLS	OBS		2N3993	0
TIS14	NJD	B15	30V	15MA	200C	0.3WF	6.5MXV	0.5/15MA	1/7.5MO		8P	4P	ALG	TIB	BC264D	2N5459	0
TIS14	NJD	B22	30V	15MA	175C	0.3WF		15MXMA	7/25MO		8P		ALN	TIW	BF810	2N6451	0
TIS25	NJD	B26	50V	10MA	175C	0.3WF	6MXV	0.5/8MA	1.5/6MO	500R	8P	4P	DUA	TIB	BFQ10	2N5045	0
TIS25X2	NJD	B26	50V	10MA	175C	0.3WF	6MXV	0.5/8MA	1.5/6MO	500R	8P	4P	DUA	TIB	BFQ10	2N5045	0
TIS26	NJD	B26	50V	10MA	175C	0.3WF	6MXV	0.5/8MA	1.5/6MO	500R	8P	4P	DUA	TIB	BFQ10	2N5045	0
TIS26X2	NJD	B26	50V	10MA	175C	0.3WF	6MXV	0.5/8MA	1.5/6MO	500R	8P	4P	DUA	TIB	BFQ10	2N5045	0
TIS27	NJD	B26	50V	10MA	175C	0.3WF	6MXV	0.5/8MA	1.5/6MO	500R	8P	4P	DUA	TIB	BFQ10	2N5045	0
TIS27X2	NJD	B26	50V	10MA	175C	0.3WF	6MXV	0.5/8MA	1.5/6MO	500R	8P	4P	DUA	TIB	BFQ10	2N5045	0
TIS33	NJD	B15	30V	50MA	175C	0.3WF	10MXV	25MNMA	12MNMO	60R	20P	5P	AMC	TIB	BSV80	2N4857	0
TIS34	NJD	B1	30V	20MA	125C	0.2WF	1/8V	4/20MA	3.5/6.5MO		6P	2P	FVG	TIB	BFW61	2N5248	0
TIS41	NJD	B6	30V	50MA	200C	0.3WF	10MXV	50MNMA		25R	18P	8P	ALC	TIB	BSV78	2N4859	0
TIS42	NJD	B1	25V	50MA	150C	.25WF	10MXV	10MNMA	20TPMO	70R	18P	9P	RLS	TIB	BSV80	2N4858	0
TIS58	NJD	B1	25V	25MA	150C	0.2WF	0.5/5V	2.5/8MA	1.3/4MO		6P	3P	ALG	TIB	BFW61	2N5458	0

TYPE NO	CONS TRUC TION	CASE & LEAD	V DS MAX	I D MAX	T J MAX	P TOT MAX	VP OR VT	IDSS OR IDOM	G MO	R DS MAX	C ISS MAX	C RSS MAX	USE	SUPP LIER	EUR SUB	USA SUB	ISS
TIS58G	NJD	B1	25	V10MA	125C	0.2WF	0.5/5V	4/8MA	1.3/4MO	-	6P	3P	ALG	TIB	BFW12	2N4303	0
TIS58Y	NJD	B1	25V	10MA	125C	0.2WF	0.5/5V	2.5/5MA	1.3/4MO		6P	3P	ALG	TIB	BFW12	2N5457	0
TIS59	NJD	B1	25V	25MA	125C	0.2WF	1/9V	6/25MA	2.3/5MO	-	6P	3P	ALG	TIB	BF244	2N5459	0
TIS59G	NJD	B1	25V	25MA	125C	0.2WF	1/9V	10/25MA	2.3/5MO	-	6P	3P	ALG	TIB	BF244	2N5459	0
TIS59Y	NJD	B1	25V	25MA	125C	0.2WF	1/9V	6/15MA	2.3/5MO	-	6P	3P	ALG	TIB	BFW61	2N5459	0
TIS68	NJD	B1	25V	10MA	125C	.36WF	0.5/2.5V	0.5/2MA	1/7.5MO		8P	4P	MPP	TIB	BFS21A		0
TIS69	NJD	B1	25V	10MA	125C	.36WF	0.5/2.5V	0.5/2MA	1/7.5MO	-	8P	4P	MPP	TIB	BFS21A		0
TIS70	NJD	B1	25V	10MA	125C	.36WF	0.5/5V	0.5/2MA	1/7.5MO	-	8P	4P	MPP	TIB	BFS21A		0
TIS73	NJD	B8	30V	100MA	150C	.36WF	4/10V	50MNMA	-	25R	10P	4P	ALC	TIB	BSV78	2N4859	0
TIS73L	NJD	B1	30V	100MA	150C	.36WF	4/10V	50MNMA	-	25R	10P	4P	ALC	TIB	BSV78	2N4859	0
TIS74	NJD	B8	30V	100MA	150C	.36WF	2/6V	20/100MA		40R	10P	3P5	ALC	TIB	BSV79	2N4860	0
TIS74L	NJD	B1	30V	100MA	150C	.36WF	2/6V	20/100MA	-	40R	10P	3P5	ALC	TIB	BSV79	2N4860	0
TIS75	NJD	B8	30V	100MA	150C	.36WF	0.8/4V	8/80MA	-	60R	10P	3P5	ALC	TIB	BSV80	2N4861	0
TIS75L	NJD	B1	30V	100MA	150C	.36WF	0.8/4V	8/80MA		60R	10P	3P5	ALC	TIB	BSV80	2N4861	0
TIS78	NJD	B8	300V	10MA	150C	.36WF	10MXV	2/10MA	0.75/3MO	1K5	15P	3P	ALH	TIB		2N6449	0
TIS79	NJD	B8	200V	10MA	150C	.36WF	12MXV	2/10MA	0.75/3MO	2K	18P	3P	ALH	TIB		2N5543	0
TIS88	NJD	B36	25V	40MA	150C	.35WF	5MXV	3/15MA	4/7MO		4P5	1P	FVG	TIB	BF256	2N5245	0
TIS88A	NJD	B1	30V	50MA	150C	.36WF	1/6V	5/15MA	4.5/7.5MO		4P5	1P	FVG	TIB	BF256	2N4416	0
TIS130	NJD	B8	30V	40MA	150C	.36WF	6MXV	40MXMA	6/10MO		15P		ALN	TIB	BF810	2N5397	0
TIXM12GE	PJD	B4	20V	25MA	125C	0.1WF	1/3.5V	5/25MA	5/20MO		15P	4P	FVG	TIB	BFT11	2W3382	0
TIXM301GE	PJD	B12	20V		100C	.15WF		25MXMA	6.5/20MO				ALG	TIB	BF320C	2N3382	0
TIXS11	PME	B12	30V		175C	0.3WF			0.8MNMO				RLS	OBS		3N156A	0
TIXS33	NJD	B15	30V	50MA	175C	0.3WF	10MXV	25MNMA	12MNMO	60R	20P	5P	AMC	TIB	BSV80	2N4857	0
TIXS35	NJD	B21	30V	200MA	175C	0.5WF	5MXV	100MXMA	10/20MO	50R	12P		RMS	TIB	BSV79	2N4857	0
TIXS36	NJD	B21	30V	200MA	175C	0.5WF	10MXV	200MXMA	10/20MO	30R	12P		RMS	TIB	BSV78	2N4391	0

TYPE NO	CONS TRUC TION	CASE & LEAD	V DS MAX	I D MAX	T J MAX	P TOT MAX	VP OR VT	IDSS OR IDOM	G MO	R DS MAX	C ISS MAX	C RSS MAX	USE	SUPP LIER	EUR SUB	USA SUB	ISS
TIXS41	NJD	B6	30V	200MA	175C	.35WF		40MNMA					RLS	OBS	BSV78	2N4859	0
TIXS42	NJD	B1	30V	200MA	125C	0.3WF		40MXMA		70R	18P	9P	RLS	OBS	BSV80	2N4861	0
TIXS59	NJD	B1	25V	25MA	125C	0.2WF	1/9V	6/25MA	2.3/5MO		6P	3P	ALG	TIB	BF244	2N5459	0
TIXS67	PME	B8	25V	120MA	150C	.36WF	1.5/5.4V	10/120MA	4MNMO		10P	4P	ALG	TIB		2N5548	0
TIXS68	NJD	B1	25V	10MA	125C	.36WF	0.5/2.5V	0.5/2MA	1/7.5MO		8P	4P	MPP	TIB	BFS21A		0
TIXS78	NJD	B8	300V	10MA	150C	.36WF	2/10V	2/10MA	.75/3MO	1.5K	15P	3P	ALH	TIB		2N5543	0
TIXS79	NJD	B8	200V	10MA	150C	.36WF	2/12V	2/10MA	0.75/3MO	2K	18P	3P	ALH	TIB		2N5543	0
TIXS80	NJD	B57	30V	75MA	175C	0.3WF	5MXV	25MXMA	5/10MO		6P	0P8	FVG	TIB		3N126	0
TIXS81	NJD	B57	30V	75MA	175C	0.3WF	10MXV	75MXMA	5/10MO		6P	0P8	FVG	TIB		3N126	0
TIX67	PME	B8	25V	125MA	150C	.36WF	5MXV	40MNMA	4MNMO		10P		ALG	OBS		3N161	0
TIX880GE	PJD	B21	40V		75C	.15WF			0.25MNMO				ALG	OBS	BF320A	2N2606	0
TIX881GE	PJD	B21	40V		75C	.15WF			0.4MNMO				ALG	OBS	BF320A	2N5265	0
TIX882GE	PJD	B21	40V		75C	.15WF			0.6MNMO				ALG	OBS	BF320A	2N2843	0
TIX883GE	PJD	B21	40V		75C	.15WF			0.8MNMO				ALG	OBS	BF320A	2N5265	0
TN4117	NJD	B15	40V	1MA	175C	.35WF	1.8MXV	0.09MXMA	0.07MNMO		3P	1P5	ALN	TDY		2N4117	0
TN4117A	NJD	B15	40V	1MA	175C	.35WF	1.8MXV	0.09MXMA	0.07MNMO		3P	1P5	ALN	TDY		2N4117A	0
TN4118	NJD	B15	40V	1MA	175C	.35WF	3MXV	0.24MXMA	0.08MNMO		3P	1P5	ALN	TDY	BF800	2N4118	0
TN4118A	NJD	B15	40V	1MA	175C	.35WF	3MXV	0.24MXMA	0.08MNMO		3P	1P5	ALN	TDY	BF800	2N4118A	0
TN4119	NJD	B15	40V	1MA	175C	.35WF	6MXV	0.2/.6MA	0.1MNMO		3P	1P5	ALN	TDY	BF800	2N4119	0
TN4119A	NJD	B15	40V	1MA	175C	.35WF	6MXV	0.2/.6MA	0.1MNMO		3P	1P5	ALN	TDY		2N4119A	0
TOR45	NJD	B59	30V	25MA	150C	.25WF	0.5/8V	2/25MA	0.3/6.5MO		8P	2P	FVG	TIB	BF245		0
TOR45A	NJD	B59	30V	25MA	150		0.4/2.2V	2/6.5MA	6.5MXMO		8P	2P	FVG	TIB	BF245		0
TOR45B	NJD	B59	30V	25MA	150C	.25WF	1.6/3.8V	6/15MA	6.5MXMA		8P	2P	FVG	TIB	BF245		0
TOR45C	NJD	B59	30V	25MA	150C	.25WF	3.2/7.5V	12/25MA	6.5MXMO		8P	2P	FVG	TIB	BF245		0
TP5114	PJD	B7	30V	90MA	200C	0.3WF	5/10V	30/90MA		75R	45P	10P	RLS	TDY		2N5114	0

TYPE NO	CONS TRUC TION	CASE & LEAD	V DS MAX	I D MAX	T J MAX	P TOT MAX	VP OR VT	IDSS OR IDOM	G MO	R DS MAX	C ISS MAX	C RSS MAX	USE	SUPP LIER	EUR SUB	USA SUB	ISS
TP5115	PJD	B7	30V	60MA	200C	0.3WF	3/6V	15/60MA		100R	45P	10P	RLS	TDY		2N5115	0
TP5116	PJD	B7	30V	25MA	200C	0.3WF	1/4V	5/25MA		150R	45P	10P	RLS	TDY		2N5116	0
U110	PJD	B7	20V	9MA	175C	0.3WF	1/6V	0.1/1MA	0.11MO		6P		ALG	SIU	BF320A	2N2606	0
U112	PJD	B7	20V	9MA	175C	0.3WF	1/6V	0.9/9MA	1MNMO		17P		ALG	SIU	BF320A	2N2608	0
U114	PJD	B7	30V	50MA	175C	0.3WF	6MXV	0.5MXMA	.11/.18MO		6P		ALG	OBS		2N2606	0
U133	PJD	B7	50V	50MA	175C	0.3WF	5MXV	1.5MXMA	.33/.53MO		10P		ALG	OBS	BF320A	2N2607	0
U139	PJD	B27	30V	50MA	200C	0.3WF	7MXV	35MXMA	7MNMO	120R	16P		DUA	OBS		2N5505	0
U139D	PJD	B27	20V	50MA	200C	0.3WF	10MXV	50MXMA	5MNMO	110R	16P		DUA	OBS		2N5505	0
U146	PJD	B7	20V	9MA	175C	0.3WF	6MXV	.025MNMA	0.0LMNMO		6P		ALG	SIU		2N2606	0
U147	PJD	B7	20V	9MA	175C	0.3WF	6MXV	.065MNMA	0.18MNMO		10P		ALG	SIU		2N2607	0
U149	PJD	B7	20V	9MA	175C	0.3WF	6MXV	0.44MNMA	1.4MNMO		30P		ALG	SIU	BF320A	IN5266	0
U168	PJD	B7	20V	6MA	150C	0.3WF	5MXV	0.6/6MA	0.8MNMO		65P		ALN	SIU	BF320A	2N5266	0
U182	NJD	B15	40V		200C	1.8WC	10MXV	120MXMA		40R	20P		RLS	OBS	BSV78	2N4091	0
U183	NJD	B15	25V	20MA	125C	0.2WF	8MXV	2/20MA	2/6.5MO		8P	4P	RLG	SIU	BC264C	2N3684	0
U184	PJD	B7	20V	9MA	175C	0.3WF	6MXV	0.2MNMA	0.54MNMO		17P		ALG	SIU	BF320A	2M2843	0
U184	NJD	B15	25V	30MA	200C	0.3WF	8MXV	3/30MA	3/8.5MO		4P	1P	RLG	SIU	BF256A	2N4416	0
U197	NJD	B6	30V	20MA	175C	0.3WF	0.2/1V	0.1/1MA	0.2MNMO	4K	7P		ALN	SIU	BF800	2N3687A	0
U198	NJD	B6	30V	20MA	175C	0.3WF	0.8/4V	0.6/6MA	0.6MNMO	1K5	7P		ALN	SIU	BF808	2N3685A	0
U199	NJD	B6	30V	20MA	175C	0.3WF	3/10V	3/20MA	1.5MNMO	650R	7P		ALN	SIU	BC264	2N3684A	0
U1277	NJD	B15	50V	10MA	200C	0.3WF	8MXV	8MXMA	0.45MNMO		6P		ALG	OBS	BC264C	2N3684	0
U1278	NJD	B15	50V	10MA	200C	0.3WF	4.5MXV	3MXMA	0.35MNMO		6P		ALG	OBS	BF347	2N3685	0
U1279	NJD	B15	50V	10MA	200C	0.3WF	2.5MXV	1.5MXMA	0.25MNMO		6P		ALG	OBS	BFW13	2N3686	0
U1279	NJD	B15	50V	10MA	200C	0.3WF	2.5MXV	1.5MXMA	0.25MNMO		6P		ALG	OBS	BFW13	2N3686	0
U1280	NJD	B15	50V	10MA	200C	0.3WF	10MXV	10MXMA	0.25MNMO		6P		ALG	OBS	BF347	2N3684	0
U1281	NJD	B6	50V	40MA	200C	0.3WF	8MXV	8MXMA	3MNMO	300R	18P		RLS	OBS		2N3966	0

TYPE NO	CONS TRUC TION	CASE & LEAD	V DS MAX	I D MAX	T J MAX	P TOT MAX	VP OR VT	IDSS OR IDOM	G MO	R DS MAX	C ISS MAX	C RSS MAX	USE	SUPP LIER	EUR SUB	USA SUB	ISS
U1282	NJD	B6	50V	40MA	200C	0.3WF	4.5MXV	20MXMA	2.5MNMO		18P		RLS	OBS		2N3824	0
U1283	NJD	B6	50V	40MA	200C	0.3WF	2.5MXV	10MXMA	1.5MNMO		18P		RLS	OBS		2N3966	0
U1284	NJD	B6	50V	40MA	200C	0.3WF	10MXV	40MXMA	1MNMO		18P		RLS	OBS	BSV80	2N4393	0
U1285	NJD	B15	30V	5MA	200C	0.3WF	8MXV	0.1MNMA	0.2/1.2MO		20P		ALG	OBS	BF800	2N3687	0
U1286	NJD	B6	30V	5MA	200C	0.3WF	8MXV	0.2MNMA	1/10MO		8P		ALG	OBS	BFW13	2N3821	0
U1287	NJD	B22	30V		200C	0.4WF	15MXV			50R	20P		RLS	OBS	BSV79	2N4860	0
U1325	NJD	B6	30V	5MA	200C	0.3WF	1.2MXV	0.5MXMA	.85/1.2MO		6P		ALG	OBS	BFW13	2N3687	0
U1714	NJD	B15	25V	5MA	200C	0.3WF	5MXV	0.5/5MA	0.4MNMO		3P	1P5	ALN	TDY	BF800	2N4340	0
U1715	NJD	B22	200V	50MA	200C	0.8WF	15MXV	10/50MA		400R	25P	4P	ALH	TDY		2N5543	0
U1837E	NJD	B10	30V	25MA	125C	0.3WF	0.5/8V	4/25MA	4.5/10MO		6P	2P	TUG	TDY	BF256C	2N5247	0
U1897E	NJD	B10	40V	150MA	125C	0.3WF	5/10V	30MNMA		30R	16P	5P	RLS	TDY	BSV78	2N4091	0
U1898E	NJD	B10	40V	100MA	125C	0.3WF	2/7V	15MNMA		40R	16P	5P	RLS	TDY	BSV79	2N4092	0
U1899E	NJD	B10	40V	80M	125C	0.3WF	1/5V	8MNMA		80R	16P	5P	RLS	TDY	BSV80	2N4093	0
U1994E	NJD	B10	30V	15MA	125C	.25WF	6MXV	5/15MA4	.5/7.5MO		4P	1P	ULG	TDY	BF256	2N5245	0
U200	NJD	B6	30V	150MA	200C	1.8WC	0.5/3V	3/25MA		150R	30P	8P	RLS	SIU		2N5555	0
U201	NJD	B6	30V	150MA	200C	1.8WC	1.5/5V	15/75MA		75R	30P	8P	RLS	SIU	BSV80	2N4092	0
U202	NJD	B6	30V	150MA	200C	1.8WC	3.5/10V	30/150MA		50R	30P	8P	RLS	SIU	BSV79	2N4091	0
U221	NJD	B22	50V	300MA	200C	0.8WF	3.5/8V	50/150MA	15MNMO		28P	7P	AMG	SIU		2N4857	0
U222	NJD	B22	50V	300MA	200C	0.8WF	6/10V	300MXMA	20MNMO		28P	7P	AMG	SIU		2N4856	0
U231	NJD	B20	50V	5MA	200C	0.3WF	0.3/4V	0.5/5MA	1/3MO		6P	2P	DUA	SIU	BF010	2N5452	0
U231X2	NJD	B20	50V	5MA	200C	0.3WF	0.3/4V	0.5/5MA	1/3MO		6P	2P	DUA	SIU	BFQ10	2N5452	0
U232	NJD	B20	50V	5MA	200C	0.3WF	0.3/4V	0.5/5MA	1/3MO		6P	2P	DUA	SIU	BFQ10	2N5452	0
U232X2	NJD	B20	50V	5MA	200C	0.3WF	0.3/4V	0.5/5MA	1/3MO		6P	2P	DUA	SIU	BFQ10	2N5452	0
U233	NJD	B20	50V	5MA	200C	0.3WF	0.3/4V	0.5/5MA	1/3MO		6P	2P	DUA	SIU	BFQ10	2N5452	0
U233X2	NJD	B20	50V	5MA	200C	0.3WF	0.3/4V	0.5/5MA	1/3MO		6P	2P	DUA	SIU	BFQ10	2N5452	0

TYPE NO	CONS TRUC TION	CASE & LEAD	V DS MAX	I D MAX	T J MAX	P TOT MAX	VP OR VT	IDSS OR IDOM	G MO	R DS MAX	C ISS MAX	C RSS MAX	USE	SUPP LIER	EUR SUB	USA SUB	ISS
U234	NJD	B20	50V	5MA	200C	0.3WF	0.3/4V	0.5/5MA	1/3MO		6P	2P	DUA	SIU	BFQ10	2N5452	0
U234X2	NJD	B20	50V	5MA	200C	0.3WF	0.3/4V	0.5/5MA	1/3MO		6P	2P	DUA	SIU	BFQ10	2N5452	0
U235	NJD	B20	50V	5MA	200C	0.3WF	0.3/4V	0.5/5MA	1/3MO		6P	2P	DUA	SIU	BFQ10	2N5452	0
U235X2	NJD	B20	50V	5MA	200C	0.3WF	0.3/4V	0.5/5MA	1/3MO		6P	2P	DUA	SIU	BFQ10	2N5452	0
U240	NJD	B6	25V	500MA	200C	0.4WF	2/10V	150MNMA		5R	70P	35P	RMS	TDY		2N5432	0
U241	NJD	B6	25V	500MA	200C	0.4WF	2/10V	100MNMA		10R	70P	35P	RMS	TDY		2N5433	0
U242	NJD	B6	20V	500MA	200C	0.4WF	2/10V	150MNMA		6R	70P	35P	RMS	TDY		2N5432	0
U243	NJD	B6	20V	500MA	200C	0.4WF	10MXV	100MNMA		12R	70P	35P	RMS	TDY		2N5433	0
U244	NJD	B67	25V	900MA	150C	10WC	3.5/8V	0.3/0.9A	80/200MO	10R	35P	15P	AHP	SIU			0
U248	NJD	B41	40V	1MA	175C	.25WF	4.5MXV	0.5MXMA	.07/.27MO		3P		DUA	OBS		2N5902	0
U248A	NJD	B41	40V	1MA	175C	.25WF	4.5MXV	0.5MXMA	.07/.27MO		3P		DUA	OBS		2N5906	0
U249	NJD	B41	40V	1MA	175C	.25WF	4.5MXV	0.5MXMA	.07/.27MO		3P		DUA	OBS		2N5903	0
U249A	NJD	B41	40V	1MA	175C	.25WF	4.5MXV	0.5MXMA	.07/.27MO		3P		DUA	OBS		2N5907	0
U250	NJD	B41	40V	1MA	175C	.25WF	4.5MXV	0.5MXMA	.07/.27MO		3P		DUA	OBS		2N5904	0
U250A	NJD	B41	40V	1MA	175C	.25WF	4.5MXV	0.5MXMA	.07/.27MO		3P		DUA	OBS		2N5908	0
U251	NJD	B41	40V	1MA	175C	.25WF	4.5MXV	0.5MXMA	.07/.27MO		3P		DUA	OBS		2N5905	0
U251A	NJD	B41	40V	1MA	175C	.25WF	4.5MXV	0.5MXMA	.07/.27MO		3P		DUA	OBS		2N5909	0
U252	NJD	B41	25V	50MA	150C	0.5WF	5MXV		5/10MO		5P		DUA	OBS		2N5911	0
U253	NJD	B41	25V	50MA	150C	0.5WF	5MXV		5/10MO		5P		DUA	OBS		2N5912	0
U254	NJD	B6	30V	250MA	200C	.36WF	10MXV	50MNMA		25R	18P		RMS	NAT	BSV78	2N4859	0
U255	NJD	B6	30V	250MA	200C	.36WF	6MXV	50MNMA		40R	18P		RMS	NAT	BSV79	2N4860	0
U256	NJD	B6	30V	250MA	200C	.36WF	4MXV	50MNMA		60R	18P		RMS	NAT	BSV80	2N4861	0
U257	NJD	B43	25V	40MA		0.5WF	1/5V	5/40MA	5/10MO		5P	1P2	DUA	INB	BFQ10	2N5564	0
U257/D	NJD	B73	25V	40MA		0.5WF	1/5V	5/40MA	5/10MO		5P	1P2	DUA	INB	BFQ10	2N5564	0
U257/W	NJD	B74	25V	40MA		0.5WF	1/5V	5/40MA	5/10MO		5P	1P2	DUA	INB	BFQ10	2N5564	0

TYPE NO	CONS TRUC TION	CASE & LEAD	V DS MAX	I D MAX	T J MAX	P TOT MAX	VP OR VT	IDSS OR IDOM	G MO	R DS MAX	C ISS MAX	C RSS MAX	USE	SUPP LIER	EUR SUB	USA SUB	ISS
U257CHP	NJD	B73	25V	40MA	200C		1/5V	5/40MA	5/10MO		5P	1P2	DUA	SIU		2N5564	0
U266	NJD	B67	20V	500MA	150	10WC	15MXV	300MA	20/40MO		28P		AMP	OBS		U244	0
U273	NJD	B15	30V	7MA	150C	0.3WF	3MXV	2MXMA	0.5MNMO		2P		ALN	OBS		2N4117	0
U273	NJD	B15	30V	7MA	150C	0.3WF	3MXV	2MXMA	0.5MNMO		2P		ALN	OBS		2N4117	0
U274	NJD	B15	30V	7MA	150C	0.3WF	5MXV	4MXMA	0.6MNMO		2P		ALN	OBS		2N4118	0
U275	NJD	B15	30V	7MA	150C	0.3WF	7MXV	6.5MXMA	0.8MNMO		2P		ALN	OBS		2N4119	0
U280	NJD	B51	50V	6MA	150C	0.5WF	4.5MXV	6MXMA	1/3MO		6P		DUA	NAT	BFQ10	2N5452	0
U280X2	NJD	B51	50V	6MA	150C	0.5WF	4.5MXV	6MXMA	1/3MO		6P		DUA	NAT	BFQ10	2N5452	0
U281	NJD	B51	50V	6MA	150C	0.5WF	4.5MXV	6MXMA	1/3MO		6P		DUA	NAT	BFQ10	2N5453	0
U281X2	NJD	B51	50V	6MA	150C	0.5WF	4.5MXV	6MXMA	1/3MO		6P		DUA	NAT	BFQ10	2N5453	0
U282	NJD	B51	50V	6MA	150C	0.5WF	4.5MXV	6MXMA	1/3MO		6P		DUA	NAT	BFQ10	2N5453	0
U282X2	NJD	B51	50V	6MA	150C	0.5WF	4.5MXV	6MXMA	1/3MO		6P		DUA	NAT	BFQ10	2N5453	0
U283	NJD	B51	50V	6MA	150C	0.5WF	4.5MXV	6MXMA	1/3MO		6P		DUA	NAT	BFQ10	2N5453	0
U283X2	NJD	B51	50V	6MA	150C	0.5WF	4.5MXV	6MXMA	1/3MO		6P		DUA	NAT	BFQ10	2N5453	0
U284	NJD	B51	50V	6MA	150C	0.5WF	4.5MXV	6MXMA	1/3MO		6P		DUA	NAT	BFQ10	2N5454	0
U284X2	NJD	B51	50V	6MA	150C	0.5WF	4.5MXV	6MXMA	1/3MO		6P		DUA	NAT	BFQ10	2N5454	0
U285	NJD	B51	50V	6MA	150C	0.5WF	4.5MXV	6MXMA	1/3MO		6P		DUA	NAT	BFQ10	2N5454	0
U285X2	NJD	B51	50V	6MA	150C	0.5WF	4.5MXV	6MXMA	1/3MO		6P		DUA	NAT	BFQ10	2N5454	0
U290	NJD	B6	30V	1.5A	200C	0.3WF	4/10V	500MNMA		2R5	60P		RHS	SIU		2N6568	0
U290CHP	NJD	B73	30V	1.5A	200C		4/10V	500MNMA		2R5	60P		RMS	SIU			0
U291	NJD	B6	30V	1.5A	200C	0.3WF	1.5/4.5V	200MNMA		7R	60P		RHS	SIU		2N6568	0
U291CHP	NJD	B73	30V	1.5A	200C		1.5/4.5V	200MNMA		7R	60P		RMS	SIU			0
U2047E	NJD	B10	30V	25MA	125C	.25WF	8MXV	4/25MA	4.5MNMO		4P	1P3	TUG	TDY	BF256C	2N5247	0
U300	PJD	B7	40V	90MA	150C	0.3WF	5/10V	30/90MA	8/12MO	60R	20P	5P5	RLS	SIU		2N5114	0
U300CHP	PJD	B73	40V	90MA	200C		5/10V	30/90MA	8/12MO		20P	5P5	ALG	SIU			0

TYPE NO	CONS TRUC TION	CASE & LEAD	V DS MAX	I D MAX	T J MAX	P TOT MAX	VP OR VT	IDSS OR IDOM	G MO	R DS MAX	C ISS MAX	C RSS MAX	USE	SUPP LIER	EUR SUB	USA SUB	ISS
U301	PJD	B7	40V	90MA	150C	0.3WF	2.5/6V	15/60MA	7/11MO	100R	20P	5P5	ALG	SIU		2N5115	0
U301CHP	PJD	B73	40V	90MA	200C		2.5/6V	15/60MA	0.7/11MO		20P	5P5	ALG	SIU			0
U304	PJD	B7	30V	90MA	150C	.35WF	5/10V	30/90MA		85R	27P	7P	RLS	SIU		2N5114	0
U305	PJD	B7	30V	90MA	150C	.35WF	3/6V	15/60MA		110R	27P	7P	RLS	SIU		2N5115	0
U306	PJD	B7	30V	90MA	150C	.35WF	1/4V	5/25MA		175R	27P	7P	RLS	SIU		2N5116	0
U308	NJD	B15	25V	60MA	150C	0.5WF	1/6V	12/60MA	10/20MO		7P5		TUG	INB		2N5398	0
U308/D	NJD	B73	25V	60MA	150C		1/6V	12/60MA	10/20MO		7P5		TUG	INB			0
U308/W	NJD	B74	25V	60MA	150C		1/6V	12/60MA	10/20MO		7P5		TUG	INB			0
U308CHP	NJD	B73	25V	60MA	200C		1/6V	12/60MA	10/20MO		7P5		ULG	SIU			0
U308T092	NJD	B63	25V	60MA	125C	0.3WF	1/6V	12/60MA	10/20MO		7P5		TUG	INB		2N5398	0
U309	NJD	B15	25V	60MA	150C	0.5WF	1/4V	12/30MA	10/20MO		7P5		TUG	INB		2N5397	0
U309/D	NJD	B73	25V	60MA	150C		1/4V	12/30MA	10/20MO		7P5		TUG	INB			0
U309/W	NJD	B74	25V	60MA	150C		1/4V	12/30MA	10/20MO		7P5		TUG	INB			0
U309CHP	NJD	B73	25V	60MA	200C		1/4V	12/30MA	10/20MO		7P5		ULG	SIU			0
U309T092	NJD	B63	25V	60MA	125C	0.3WF	1/4V	12/30MA	10/20MO		7P5		TUG	INB		2N5397	0
U310	NJD	B15	25V	60MA	150C	0.5WF	2.5/6V	24/60MA	10/18MO		7P5		TUG	INB			0
U310/D	NJD	B73	25V	60MA	150C		2.5/6V	24/60MA	10/18MO		7P5		TUG	INB			0
U310/W	NJD	B74	25V	60MA	150C		2.5/6V	24/60MA	10/18MO		7P5		TUG	INB			0
U310CHP	NJD	B73	25V	60MA	200C		2.5/6V	24/60MA	10/18MO		7P5		ULG	SIU			0
U310T092	NJD	B63	25V	60MA	150C	0.3WF	2.5/6V	24/60MA	10/18MO		7P5		TUG	INB			0
U311	NJD	B6	25V	60MA	150C	0.3WF	1/6V	20/60MA	10/20MO		7P5		TUG	INB			0
U312	NJD	B6	25V	30MA	150C	0.5WF	1/6V	10/30MA	6/10MO		5P		RLG	SIU	BF348	2N5398	0
U314	NJD	B69	25V	30MA	200C	.17WF	1/6V	10/30MA	6/10MO		5P	1P2	ULA	SIU		2N5397	0
U315	NJD	B68	25V	30MA	200C	.17WF	1/6V	10/30MA	6/10MO		5P	1P2	ULA	SIU		2N5397	0
U316	NJD	B69	25V	60MA	200C	.17WF	1/4V	12/30MA			6P		ULA	SIU	BF256C	2N5427	0

TYPE NO	CONS TRUC TION	CASE & LEAD	V DS MAX	I D MAX	T J MAX	P TOT MAX	VP OR VT	IDSS OR IDOM	G MO	R DS MAX	C ISS MAX	C RSS MAX	USE	SUPP LIER	EUR SUB	USA SUB	ISS
U317	NJD	B69	25V	60MA	200C	.17WF	2.5/6V	24/60MA			6P		ULA	SIU	BF348	2N5398	0
U320	NJD	B23	25V	700MA	150C	2WC	2/10V	0.1/.5A	.075/.2MO	10R	30P	15P	VMP	SIU			0
U321	NJD	B23	25V	700MA	150C	2WC	1/4V	80/250MA	.075/.2MO	11R	30P	15P	VMP	SIU			0
U321CHP	NJD	B73	25V	700MA	200C		1/4V	80/250MA	.075/.2MO		30P	15P	RLG	SIU			0
U322	NJD	B23	25V	700MA	150C	2WC	3/10V	0.2/.7A	.075/.2MO	8R	30P	15P	VMP	SIU		2N6568	0
U322CHP	NJD	B73	25V	700MA	200C		3/10V	0.2/.7A	.075/.2MO		30P	15P	RLG	SIU			0
U328	NJD	B22	275V	60MA	200C	0.8WF	5/15V	10/60MA	2.5/6MO	400R	25P	4P	ALH	SIU		2N5543	0
U328CHP	NJD	B73	275V	60MA	200C		5/15V	10/60MA	2/6MO	500R	25P	4P	ALH	SIU			0
U329	NJD	B22	200V	60MA	200C	0.8WF	5/15V	10/60MA	2.5/6MO	400R	25P	4P	ALH	SIU		2N5543	0
U329CHP	NJD	B73	200V	60MA	200C		5/15V	10/60MA	2/6MO	500R	25P	4P	ALH	SIU			0
U330	NJD	B22	275V	60MA	200C	0.8WF	1/7V	1/20MA	1.5/4.5MO	650R	25P	4P	ALH	SIU		2N5543	0
U330CHP	NJD	B73	275V	60MA	200C		1/7V	1/20MA	1/5MO	1K	25P	4P	ALH	SIU			0
U331	NJD	B22	200V	60MA	200C	0.8WF	1/7V	1/20MA	1.5/2.5MO	650R	25P	4P	ALH	SIU		2N5543	0
U331CHP	NJD	B73	200V		200C		1/7V	1/20MA	1/5MO	1K	25P	4P	ALH	SIU			0
U401	NJD	B20	50V	10MA	200C	0.5WF	2.5MXV	10MXMA	2/7MO		8P		DUA	SIU	BFQ10	2N5545	0
U402	NJD	B20	50V	10MA	200C	0.5WF	2.5MXV	10MXMA	2/7MO		8P		DUA	SIU	BFQ10	2N5545	0
U403	NJD	B20	50V	10MA	200C	0.5WF	2.5MXV	10MXMA	2/7MO		8P		DUA	SIU	BFQ10	2N5545	0
U404	NJD	B20	50V	10MA	200C	0.5WF	2.5MXV	10MXMA	2/7MO		8P		DUA	SIU	BFQ10	2N5545	0
U405	NJD	B20	50V	10MA	200C	0.5WF	2.5MXV	10MXMA	2/7MO		8P		DUA	SIU	BFQ10	2N5545	0
U406	NJD	B20	50V	10MA	200C	0.5WF	2.5MXV	10MXMA	2/7MO		8P		DUA	SIU	BFQ10	2N5545	0
U421	NJD	B41	60V	1MA	150C	.75WF	2MXV	1MXMA	0.3/0.8MO		3P		DUA	SIU		2N4082	0
U422	NJD	B41	60V	1MA	150C	.75WF	2MXV	1MXMA	0.3/0.8MO		3P		DUA	SIU		2N4082	0
U423	NJD	B41	60V	1MA	150C	.75WF	2MXV	1MXMA	0.3/0.8MO		3P		DUA	SIU		2N4082	0
U424	NJD	B41	60V	1MA	150C	.75WF	3MXV	1MXMA	0.3/1MO		3P		DUA	SIU		2N4082	0
U425	NJD	B41	60V	1MA	150C	.75WF	3MXV	1MXMA	0.3/1MO		3P		DUA	SIU		2N4082	0

TYPE NO	CONS TRUC TION	CASE & LEAD	V DS MAX	I D MAX	T J MAX	P TOT MAX	VP OR VT	IDSS OR IDOM	G MO	R DS MAX	C ISS MAX	C RSS MAX	USE	SUPP LIER	EUR SUB	USA SUB	ISS
U426	NJD	B41	60V	1MA	150C	.75WF	3MXV	1MXMA	0.3/1MO		3P		DUA	SIU		2N4082	0
U430	NJD	B44	25V	60MA	175C	0.3WF	1/4V	12/30MA	10/20MO		7P5		DUA	SIU		2N5911	0
U430CHP	NJD	B73	25V	60MA	200C		1/4V	12/30MA	10/20MO		7P5		DUA	SIU			0
U431	NJD	B44	25V	60MA	175C	0.3WF	2/6V	24/60MA	10/20MO		7P5		DUA	SIU		2N5911	0
U431CHP	NJD	B73	25V	60MA	200C		2/6V	20/60MA	10/20MO		7P5		DUA	SIU			0
UC20	NJD	B15	30V	20MA	200C	60MWF	5MXV	2MXMA	0.3MNMO		2P		ALG	OBS	BFW13	2N3686	0
UC21	NJD	B15	30V	5MA	200C	20MWF	2.5MXV	0.6MXMA	0.2MNMO		2P		ALG	OBS	BF800	2N3687	0
UC40	PJD	B12	30V	1MA	175C	0.2WF	5MXV	1MXMA	0.15MNMO				ALG	OBS	BF320A	2N2843	0
UC41	PJD	B12	30V	1MA	175C	0.2WF	2.5MXV	0.3MXMA	0.1MNMO		2P5		ALN	OBS	BF320A	2N2606	0
UC42	PJD	B69	30V	1MA	125C	0.2WF	2.5MXV	1MXMA					ALG	OBS		2N3575	0
UC43	PJD	B69	30V	1MA	125C	0.2WF	2.5MXV	0.3MXMA					ALG	OBS		2N3574	0
UC100	NJD	B15	30V	10MA	200C	0.3WF	5MXV	2.5/7MA	2MNMO	600R	5P		ALG	OBS	BC264B	2N3684	0
UC105	NJD	B47	30V		200C	0.3WF	5MXV	2.5/7MA	2MNMO	600R	5P		ALG	OBS	BC264B	2N3684	0
UC110	NJD	B15	30V	5MA	200C	0.3WF	3MXV	1/3MA	1.5MNMO		5P		ALG	OBS	BFW12	2N3685	0
UC115	NJD	B47	30V	3MA	200C	0.3WF	3MXV	1/3MA	1.5MNMO	800R	5P		ALG	OBS	BC264A	2N5457	0
UC120	NJD	B15	30V		200C	0.3WF	1.7MXV	.4/1.2MA	1MNMO		5P		ALG	OBS	BF800	2N3686	0
UC125	NJD	B47	30V	10MA	200C	0.3WF	1.7MXV	.4/1.2MA	1MNMO		5P		ALG	OBS	BFW13	2N3821	0
UC130	NJD	B15	30V	1MA	200C	0.3WF	1.2MXV	0.1/.5MA	0.5MNMO		5P		ALG	OBS	BF800	2N3687	0
UC135	NJD	B47	30V	1MA	200C	0.3WF	1.2MXV	0.1/.5MA	0.5MNMO		5P		ALG	OBS	BF800	2N3687	0
UC150	NJD	B15	30V		200C	0.3WF			3.5MNMO		4P		FVG	OBS		2N4416	0
UC155	NJD	B15	30V	30MA	200C	0.3WF	10MXV	10MNMA		125R	4P	1P	ULA	OBS		2N4393	0
UC155E	NJD	B15	30V	50MA	125C	0.2WF	10MXV	10MNMA					ULA	OBS	BF256	2N5247	0
UC155N	NJD	B69	30V		125C	0.2WF	10MXV	10MNMA					ULA	OBS		U314	0
UC155W	NJD	B69	30V	30MA	125C	.17WF	10MXV	10/30MA		125R	3P5	1P	ULA	OBS		U314	0
UC200	NJD	B15	50V	30MA	200C	.35WF	6MXV	10/30MA	6MNMO	150R	7P		FVG	OBS	BF348	2N5397	0

TYPE NO	CONS TRUC TION	CASE & LEAD	V DS MAX	I D MAX	T J MAX	P TOT MAX	VP OR VT	IDSS OR IDOM	G MO	R DS MAX	C ISS MAX	C RSS MAX	USE	SUPP LIER	EUR SUB	USA SUB	ISS
UC201	NJD	B15	50V	200MA	200C	.35WF	8MXV	15MNMA		125R	7P	4P	RLS	OBS	BSV80	2N5555	0
UC210	NJD	B15	50V	12MA	200C	.35WF	4MXV	4/12MA	4.5MNMO	250R	7P		ALN	OBS	BF810	2N3822	0
UC220	NJD	B15	50V	5MA	200C	.35WF	2.5MXV	1/5MA	3MNMO		7P		ALN	OBS	BFW12	2N3822	0
UC240	NJD	B6	50V	10MA	200C	0.3WF	5MXV	10MXMA	1.2MNMO		1P8		RLS	OBS	BF256A	2N4340	0
UC241	NJD	B15	50V	10MA	200C	0.3WF	5MXV	1/10MA	2MNMO		20P		ALN	OBS	BC264	2N4221	0
UC250	NJD	B6	30V	150MA	200C	0.3WF	5/10V	50/150MA		30R	25P	7P	RLS	TDY	BSV78	2N4859	0
UC251	NJD	B6	30V	75MA	200C	0.3WF	1/6V	7.5/75MA		75R	25P	7P	RLS	TDY	BSV80	2N4861	0
UC258	NJD	B15	30V		200C	0.3WF			11MNMO		14P		FVG	OBS	BF348	2N5397	0
UC300	PJD	B15	30V			0.3WF	5MXV	3.8MXMA	1MNMO	1K2	5P		ALG	OBS	BF320A	2N5267	0
UC305	PJD	B47	30V	5MA	175C	0.3WF	5MXV	3.8MXMA	1MNMO	1K2	5P		ALG	OBS	BF320A	2N5267	0
UC310	PJD	B15	30V		200C	0.3WF			0.75MNMO				ALG	OBS	BF320A	2N5265	0
UC315	PJD	B47	30V	2MA	200C	0.3WF	3MXV	1.5MXMA	0.75MNMO	1K6	5P		ALG	OBS	BF320A	2N5266	0
UC320	PJD	B15	30V		200C	0.3WF			0.5MNMO				ALG	OBS	BF320A	2N5265	0
UC325	PJD	B47	20V	1MA	200C	0.3WF	1.7MXV	0.6MXMA	0.3MNMO	2K4	5P		ALG	OBS	BF320A	2N2843	0
UC330	PJD	B15	30V		200C	0.3WF			0.25MNMO				ALG	OBS	BF320A	2N2607	0
UC335	PJD	B47	30V	1MA	200C	0.3WF	1.2MXV	0.25MXMA	0.25MNMO	4K8	5P		ALG	OBS	BF320A	2N2607	0
UC340	PJD	B15	30V		200C	0.3WF			0.33MNMO				ALG	OBS	BF320A	2N2607	0
UC400	PJD	B12	30V	15MA	200C	.35WF	6MXV	5/15MA	3MNMO		8P		ALN	OBS	BF320C	2N5462	0
UC401	PJD	B12	30V	10MA	175C	0.3WF	8MXV	8MXMA		250R	8P	4P	RLS	OBS		2N3993	0
UC410	PJD	B15	30V	10MA	200C	.18WF	4MXV	2/6MA	2.2MNMO	500R	8P		ALN	OBS	BF320B	2N5269	0
UC420	PJD	B15	30V	5MA	200C	0.2WC	2.5MXV	.5/2.5MA	1.5MNMO	700R	8P		ALN	OBS	BF320A	2N5267	0
UC450	PJD	B7	25V	75MA	200C	0.5WF	10MXV	75MXMA	10MNMO		25P		RLS	NAB		2N5115	0
UC451	PJD	B7	25V	400MA	200C	0.5WF	6MXV	375MXMA	6MNMO		25P		RLS	NAB		2N5114	0
UC588	NJD	B58	30V	15MA	125C	.36WF	6MXV	5/15MA	4.5/7.5MO		4P5	1P	TUG	OBS	BF256	2N4416	0
UC701	NJD	B15	40V	3MA	200C	0.3WF	6MXV	3MXMA					ALG	OBS	BC264A	2N3685	0

TYPE NO	CONS TRUC TION	CASE & LEAD	V DS MAX	I D MAX	T J MAX	P TOT MAX	VP OR VT	IDSS OR IDOM	G MO	R DS MAX	C ISS MAX	C RSS MAX	USE	SUPP LIER	EUR SUB	USA SUB	ISS
UC703	NJD	B15	40V	10MA	200C	0.3WF	6MXV	0.1/10MA	0.5/5MO	2K	6P		ALG	OBS	BF808	2N4220	0
UC704	NJD	B15	40V	25MA	200C	0.3WF	8MXV	0.2/24MA	1/10MO	1K	8P		ALG	OBS	BFW61	2N4220	0
UC705	NJD	B15	40V	50MA	200	0.3WF	8MXV	0.5/50MA	2/20MO		12P		ALG	OBS	BF817	2N6451	0
UC707	NJD	B6	20V	250MA	200C	1.8WC	12MXV	3/250MA	5/50MO	50R	30P		RLS	OBS	BSV79	2N4860	0
UC714	NJD	B15	30V	20MA	200C	0.3WF	8MXV	2/20MA	2/6.5MO		8P	4P	ALG	NAB	BFW61	2N3819	0
UC714E	NJD	B10	30V	20MA	125C	0.2WF	8MXV	2/20MA	4TPMO		8P	4P	ALG	OBS	BFW61	2N5486	0
UC734	NJD	B15	30V	20MA	200C	0.3WF	1/8V	4/20MA	3.5/6.5MO		4P	OP8	FVG	NAB	BFW61	2N3823	0
UC734E	NJD	B10	30V	20MA	125C	0.3WF	1/8V	4/20MA	3.5/6.5MO		4P5	1P	FVG	NAB	BFW61	2N5486	0
UC750	NJD	B47	30V	10MA	200C	0.3WF	6MXV	0.05MNMA	0.1MNMO				ALG	OBS	BF800	2N3687	0
UC751	NJD	B47	30V	10MA	200C	0.3WF	6MXV	0.1MNMA	0.35MNMO		10P		ALG	OBS	BFW13	2N3687	0
UC752	NJD	B47	30V	5MA	200C	0.3WF	6MXV	0.3MNMA	1MNMO		17P		ALG	OBS	BF800	2N3686	0
UC753	NJD	B47	30V	5MA	200C	0.3WF	6MXV	0.9MNMA	2.5MNMO		25P		ALG	OBS	BFW12	2N3684	0
UC754	NJD	B6	30V	15MA	175C	0.3WF	4MXV	0.5/5MA	1MNMO		6P	3P	ALG	OBS	BFW12	2N4340	0
UC755	NJD	B6	30V	15MA	175C	0.3WF	6MXV	4/10MA	2MNMO		6P	3P	ALG	OBS	BC264	2N4341	0
UC756	NJD	B6	30V	15MA	175C	0.3WF	10MXV	0.5/15MA	1MNMO		6P	3P	ALG	OBS	BFW61	2N4340	0
UC801	PJD	B15	25V		175C	0.3WF	6MXV	1.5MNMA					ALG	OBS	BF320A	2N5267	0
UC803	PJD	B15	25V		175C	0.3WF	6MXV	5MNMA					ALG	OBS		2N3331	0
UC804	PJD	B15	25V		175C	0.3WF	8MXV	12MNMA					ALG	OBS	BFT11	2N3384	0
UC805	PJD	B15	25V	25MA	175C	0.3WF	8MXV	0.3/25MA	1/10MO	1K	12P		ALG	OBS	BF320C	2N3331	0
UC807	PJD	B12	20V	250MA	175C	0.6WF	12MXV	1/125MA	2.5/25MO	400R	30P		RLS	OBS		2N3386	0
UC814	PJD	B15	25V	15MA	175C	0.3WF	8MXV	0.3/15MA	0.8/5MO	1K3	16P	8P	ALG	OBS	BF320C	2N3331	0
UC850	PJD	B6	20V		200C	0.3WF	6MXV	1MNMA					ALG	OBS		2N3329	0
UC851	PJD	B7	20V	10MA	200C	0.3WF	6MXV	0.9/9MA	1MNMO		17P		ALG	OBS	BF320A	2N2608	0
UC852	PJD	B6	25V		200C	0.3WF		0.02MNMA	0.06MNMO		6P		ALG	OBS	BF320A	2N2606	0
UC853	PJD	B7	25V	10MA	200C	0.3WF	6MXV	.07MNMA	0.18MNMO		10P		ALG	OBS	BF320A	2N2607	0

TYPE NO	CONS TRUC TION	CASE & LEAD	V DS MAX	I D MAX	T J MAX	P TOT MAX	VP OR VT	IDSS OR IDOM	G MO	R DS MAX	C ISS MAX	C RSS MAX	USE	SUPP LIER	EUR SUB	USA SUB	ISS
UC854	PJD	B7	25V	10MA	200C	0.3WF	6MXV	0.2MNMA	0.54MNMO		17P		ALG	OBS	BF320B	2N2608	0
UC855	PJD	B7	25V	10MA	200C	0.3WF	6MXV	.44MNMA	1.4MNMO		25P		ALG	OBS	BF320B	2N2609	0
UC1700	PME	B15	40V				2.5/5V		2/4MO	400R	5P	1P2	RLS	OBS		3N156A	0
UC1764	PME	B52	30V	50MA	200C	.37WF	5MXV	3MNMA	1/4MO	300R	3P		RLS	OBS		3N163	0
UC2130	NJD	B51	50V	5MA	200C	0.5WF	5MXV	4.5MXMA	1MNMO		4P		DUA	OBS	BFQ10	2N5452	0
UC2130X2	NJD	B51	50V	5MA	200C	0.5WF	5MXV	4.5MXMA	1MNMO		4P		DUA	OBS	BFQ10	2N5452	0
UC2132	NJD	B51	50V	5MA	200C	0.5WF	5MXV	4.5MXMA	1MNMO		4P		DUA	OBS	BFQ10	2N5452	0
UC2132X2	NJD	B51	50V	5MA	200C	0.5WF	5MXV	4.5MXMA	1MNMO		4P		DUA	OBS	BFQ10	2N5452	0
UC2134	NJD	B51	50V	5MA	200	0.5WF	5MXV	4.5MXMA	1MNMO		4P		DUA	OBS	BFQ10	2N5452	0
UC2134X2	NJD	B51	50V	5MA	200C	0.5WF	5MXV	4.5MXMA	1MNMO		4P		DUA	OBS	BFQ10	2N5452	0
UC2136	NJD	B51	50V	5MA	200C	0.5WF	5MXV	4.5MXMA	1MNMO		4P		DUA	OBS	BFQ10	2N5452	0
UC2136X2	NJD	B51	50V	5MA	200C	0.5WF	5MXV	4.5MXMA	1MNMO		4P		DUA	OBS	BFQ10	2N5452	0
UC2138	NJD	B51	50V	5MA	200C	0.5WF	5MXV	4.5MXMA	1MNMO		4P		DUA	OBS	BFQ10	2N5452	0
UC2138X2	NJD	B51	50V	5MA	200C	0.5WF	5MXV	4.5MXMA	1MNMO		4P		DUA	OBS	BFQ10	2N5452	0
UC2139	NJD	B51	30V	10MA	200C	0.5WF		6MXMA					DUA	OBS	BFQ10	2N5515	0
UC2139X2	NJD	B51	30V	10MA	200C	0.5WF		6MXMA					DUA	OBS	BFQ10	2N5515	0
UC2147	NJD	B20	30V		200C	.25WF			1MNMO	1K2	6P		DUA	OBS	BFQ10	2N5452	0
UC2147X2	NJD	B20	30V		200C	.25WF			1MNMO	1K2	6P		DUA	OBS	BFQ10	2N5452	0
UC2148	NJD	B51	50V	15MA	200C	0.3WF	6MXV	15MXMA	1MNMO				DUA	OBS	BFQ10	2N5561	0
UC2148X2	NJD	B51	50V	15MA	200C	0.3WF	6MXV	15MXMA	1MNMO				DUA	OBS	BFQ10	2N5561	0
UC2149	NJD	B51	30V	15MA	200C	0.3WF	6MXV	2MNMA	2MNMO				DUA	OBS	BFQ10	2N5561	0
UC2149X2	NJD	B51	30V	15MA	200C	0.3WF	6MXV	2MNMA	2MNMO				DUA	OBS	BFQ10	2N5561	0
UC2766	PME	B38	30V	50MA	150C	0.3WF	5MXV	30MXMA	1/4MO	300R	3P5	1P	ALG	OBS		3N166	0
UCX1702	PME	B15	30V				2/5V		1MNMO	250R	5P	1P2	RLS	OBS		3N156A	0
UT100	NJD	B68	25V	30MA	150C	0.3WF	6MXV	30MXMA6	6/10MO		5P		TUG	OBS		U315	0

TYPE NO	CONS TRUC TION	CASE & LEAD	V DS MAX	I D MAX	T J MAX	P TOT MAX	VP OR VT	IDSS OR IDOM	G MO	R DS MAX	C ISS MAX	C RSS MAX	USE	SUPP LIER	EUR SUB	USA SUB	ISS
UT101	NJD	B68	25V	30MA	150C	0.3WF	6MXV	30MXMA	6/10MO		5P		TUG	OBS		U315	0
VCR2N	NJD	B6	15V	5MA	200C	0.3WF	3.5/7V			60R	15P		VCR	SIU			0
VCR3P	PJD	B12	15V		200C	0.3WF	3.5/7V			200R	12P		VCR	SIU			0
VCR4N	NJD	B6	15V	5MA	200C	0.3WF	3.5/7V			600R	6P		VCR	SIU			0
VCR5P	PJD	B12			200C	0.3WF	3.5/7V			900R	6P		VCR	SIU			0
VCR6P	PJD	B7	15V		200C	0.3WF	2/4V			900R	50P		VCR	SIU			0
VCR7N	NJD	B14	15V	5MA	200C	0.3WF	2.5/5V			8K	50P		VCR	SIU			0
VCR10N	NJD	B15	25V		200C	0.3WF	8/12V			200R	12P		VCR	SIU			0
VCR11N	NJD	B20	25V		200C	0.3WF	8/12V			200R	12P		VCR	SIU			0
VCR20N	NJD	B6	30V		200C	0.3WF	15/25V			160R	6P		VCR	SIU			0
VF28	PJD	B50	20V	30MA	125C	0.2WF	10MXV	30MXMA	8/100MO	700R	20P		RLS	OBS		2N4392	0
VI1010	PME		50V	25MA	125C	.11WF	6MXV		0.5MNMO		1P1		RLS	OBS		3N174	0
WK5457	NJD	B3	25V	10MA	135C	0.3WF	6MXV	10MXMA	1/5MO		7P		ALG	OBS	BF347	2N5457	0
WK5458	NJD	B3	25V	10MA	135C	0.3WF	7MXV	10MXMA	1.5/5.5MO		7P		ALG	OBS	BF347	2N5458	0
WK5459	NJD	B3	25V	10MA	135C	0.3WF	8MXV	10MXMA	2/6MO		7P		ALG	OBS	BC264C	2N5459	0
ZFT12	NJD	B24	25V	1MA	150C	.35WF	2.4MXV	1MXMA	0.4/1MO				ALG	OBS	BF800	2N3687	0
ZFT12A	NJD	B24	25V	1MA	150C	.35WF	2.4MXV	1MXMA	0.4/1MO				ALG	OBS	BF800	2N3687	0
ZFT14	NJD	B24	25V	6MA	150C	.35WF	8MXV	6MXMA	0.9/2MO				ALG	OBS	BF808	2N3684	0
ZFT14A	NJD	B24	25V	6MA	150C	.35WF	8MXV	6MXMA	0.9/2MO				ALG	OBS	BF808	2N3684	0
ZFT16	NJD	B24	60V	6MA	150C	.35WF	8MXV	6MXMA	0.9/2MO				ALG	OBS	BFW12	2N4886	0
ZFT18	NJD	B24	100V	6MA	150C	.35WF	8MXV	6MXMA	0.9/2MO				ALG	OBS		2N4886	0
ZTX350	PME	B4	20V	10MA	125C	0.3WF	1/6V	5MXMA	0.5/2.5MO	1K5	4P5		ALG	FEB	BSW95	3N174	0

Appendix A
Explanatory Notes to Tabulations

The general layout plan of the information in the tables of this selector should be immediately evident from the data tabulation explanatory chart set out below.

TYPE NO	CON STRU CTION	CASE & LEADS	V_{DS} MAX	I_D MAX	T_j MAX	P_{TOT} MAX	V_P or V_T	I_{DSS} or I_{DON}
(EXAMPLE) 2N3819	NJD	B1	25V	20 MA	135C	0.3 WF	0.5 CV	1/5 MA

NUMERO-ALPHABETIC LISTING

D = DEPLETION
E = ENHANCEMENT
J = JUNCTION-GATE
M = MOSFET (IGFET)
N = N-CHANNEL
P = P-CHANNEL
X = DEPLN/ENHANCT

REFER TO CASE OUTLINES & LEADS APPENDIX B

MAXIMUM PERMISSIBLE DRAIN-SOURCE VOLTAGE

MAXIMUM PERMISSIBLE DRAIN CURRENT

MAXIMUM PERMISSIBLE JUNCTION TEMPERATURE

MAXIMUM PERMISSIBLE DEVICE DISSIPATION - "F" = FREE AIR AT 25°C; "C" = CASE SURFACE HELD AT 25°C; "H" = IN FREE AIR AT 25°C WITH METAL HEAT SINK ATTACHED TO DEVICE

V_P & V_T = "PINCH-OFF" (DEPLETION TYPES)) & "THRESHOLD" (ENHANCEMENT TYPES) VOLTAGE: I.E. CUT-OFF GATE-BIAS VOLTAGE EXPRESSED IN VOLTS (V) WITH "MX" = MAX.; "MN" = MIN; "TP" = TYPICAL AND "/" = RANGE

"ON" DRAIN CURRENT WITH GATE SHORTED TO SOURCE (DEPLETION) OR TO DRAIN (ENHANCEMENT)

G_{mo}	R_{DS} MAX	C_{ISS} MAX	C_{RSS} MAX	USE	SUP PLI ER	EUR SUB	USA SUB
1/5 MO	–	7P	3P	ALG	TIU	BF 244A	2N5457

EIA-JEDEC 2N or 3N STANDARD POSSIBLE SUBSTITUTE

PROELECTRON STANDARD POSSIBLE SUBSTITUTE

CODE INDICATION OF POSSIBLE SUPPLIER OF DEVICE – SEE MANUFACTURER LISTINGS APPENDIX D

CODE INDICATION OF APPLICATION USAGE – SEE BELOW IN THIS APPENDIX

MAXIMUM DRAIN – GATE FEEDBACK. CAPACITANCE (TYPICAL BEING USUALLY ½ TO ⅔ MAX) – SPECIFIED IN PICOFARADS (P)

MAXIMUM GATE INPUT CAPACITANCE TYPICAL BEING ½-⅔ MAX) SPECIFIED IN PICOFARADS (P)

MAXIMUM DRAIN – SOURCE "ON" RESISTANCE – SPECIFIED IN OHMS (R)

TRANSCONDUCTANCE AT MAXIMUM BIAS CURRENT – SPECIFIED IN MILLIMHOS (MO).

More detailed notes on the symbols and codings used are given below.

Column headings

$V_{DS\ max}$	Drain-source voltage rating
$I_{D\ max}$	Rated maximum drain current
$T_{j\ max}$	Rated maximum junction temperature
$P_{TOT\ max}$	Rated maximum device power dissipation
Vp/VT	Pinch-off (Vp) or threshold (VT) voltage
I_{DSS}/I_{DON}	Drain saturation current
Gme	Transconductance at saturation drain current
R_{DS}	Drain source resistance at saturation
C_{ISS}	Imput capacitance at gate
C_{RSS}	Feedback capacitance at drain

Units

A = Amperes
C = ° Centigrade
K = Kilohertz
M = Megahertz
MA = Milliamperes
MN = Minimum
MWC = Milliwatts, case at 25° C
MWF = Milliwatts, free air 25°C
MWH = Milliwatts, heat sink, 25°C ambient
P = Picofarads (refers C_{oss} & C_{rss})
TP = Typical
UA = Microampere
V = Volts
WC = Watts, case at 25°C
WF = Watts, free air 25°C
WH = Watts, heat sink, 25°C ambient

Where unit appears in the middle of the value it indicates the position of the decimal point; e.g. 3P5 = 3.5P = 3.5 picofarads)

'Case & Lead Identification' column codes

B . . . = number of related drawing in Appendix B. ('OBS' = obsolete).

'Use' column codes

In the three letter code used in this column to indicate applications use, different systems are used for industrial, consumer and special applications as follows:

1. Industrial . . . first letter A,R,S,U or V; each three letters significant as follows:

(First Letter)	(Second Letter)	(Third Letter)
A = Audio	H = High current	A = Amplifier
R = Rf	L = Low current	B = Bidirectional
S = Shf	M = Medium current	C = Chopper
U = Uhf		E = Extra high voltage
V = Vhf		G = General purpose
		H = High voltage
		N = Low noise
		S = Switch

2. Consumer . . . first letter F or T; code significance as follows:

FRH = Fm/am radio, general purpose, high gain
FRM = Fm/am radio, general purpose, medium gain
FVG = Fm and Vhf (TV), general purpose
TIA = TV, if amplifier
TIG = TV, if amplifier, gain controlled
TLH = TV, horizontal (line) output, high voltage
TLM = TV, horizontal (line) output, medium voltage
TLE = TV, horizontal (line) output, extra high voltage
TUG = TV, uhf amplifier, gain controlled
TUM = TV, uhf mixer
TUO = TV, uhf oscillator
TVE = TV, video output, extra high voltage
TVH = TV, video output, high voltage
TVM = TV, video output, medium voltage

3. Special

DUA = Dual or differential amplifier pair
MPP = Matched pair
PHT = Photo-device

'Supplier' column codes

A three-letter code indicating a supplier of the related device according to the details given in Appendix D. ('OBS' implies obsolete type).

Appendix B

Package Outline and Lead/Terminal Diagrams

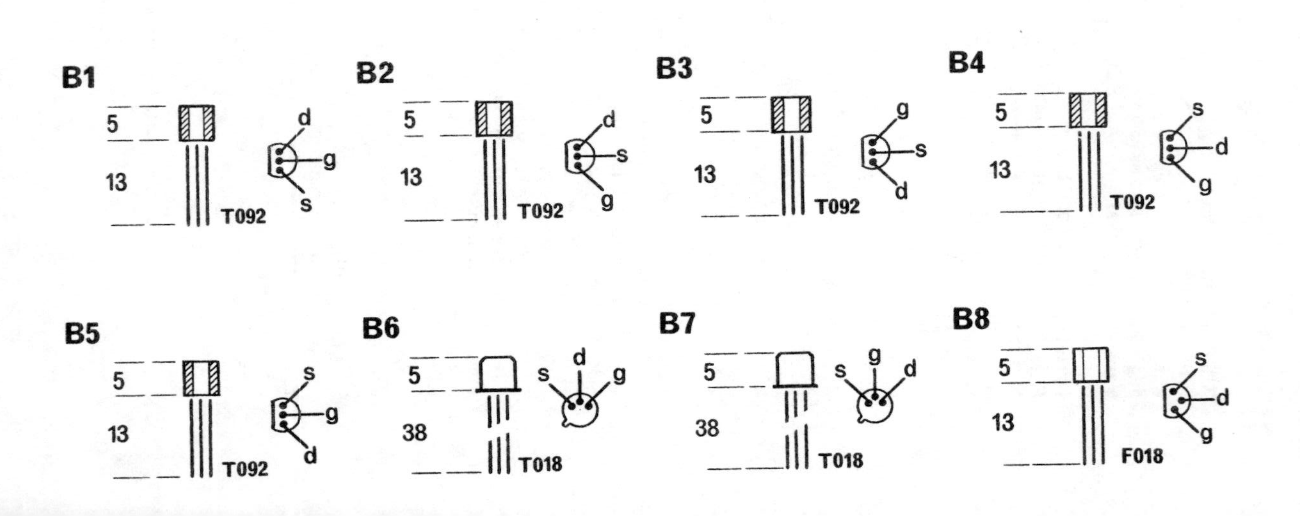

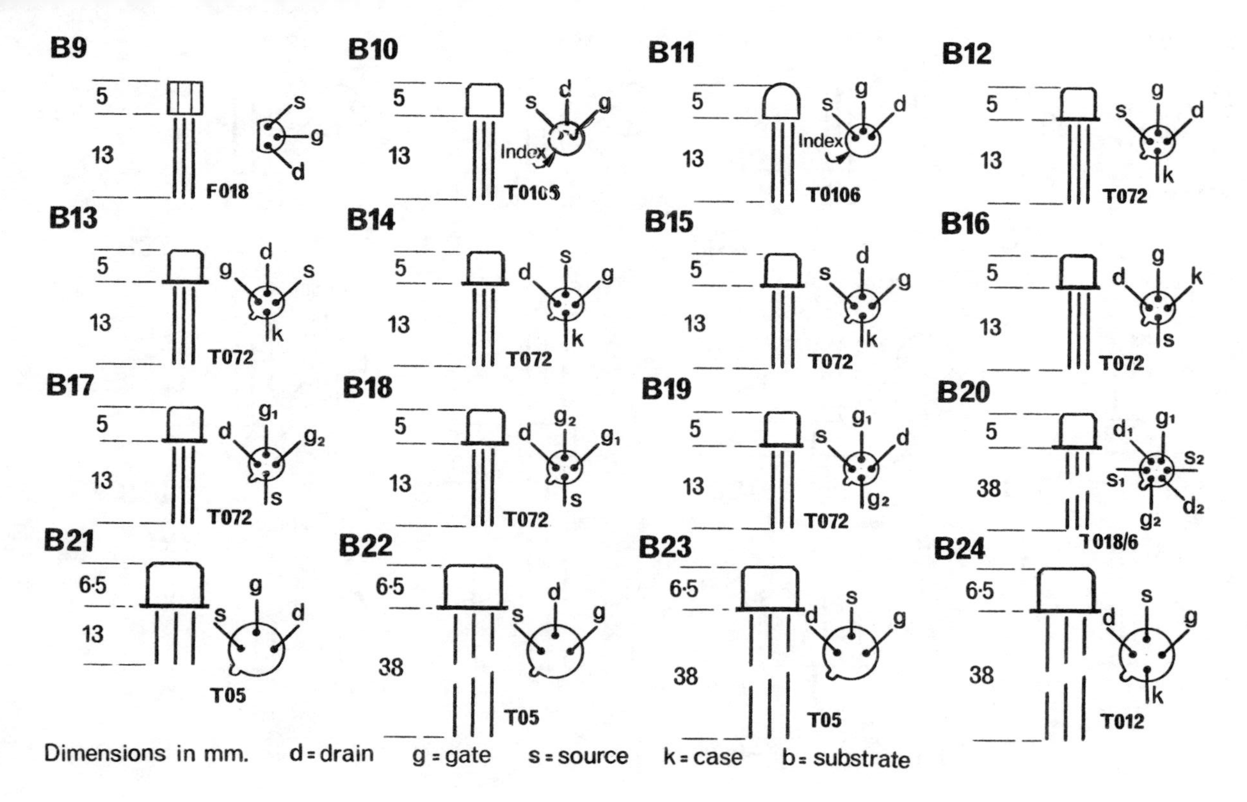

Dimensions in mm. d = drain g = gate s = source k = case b = substrate

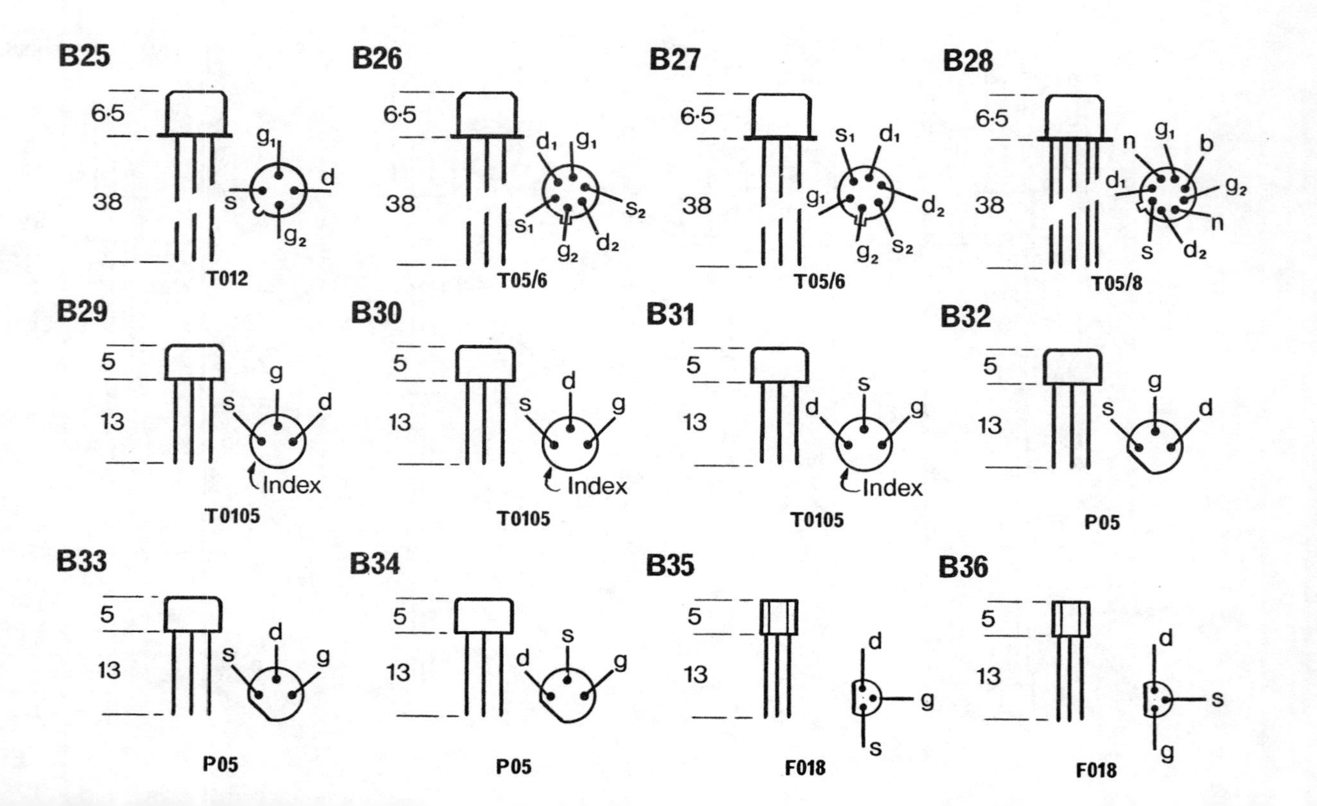
B25
6·5
38
g_1
d
s
g_2
T012
B26
6·5
38
d_1
g_1
s_1
s_2
g_2
d_2
T05/6
B27
6·5
38
s_1
d_1
g_1
d_2
g_2
s_2
T05/6
B28
6·5
38
n
g_1
b
d_1
g_2
n
s
d_2
T05/8
B29
5
13
g
s
d
Index
T0105
B30
5
13
d
s
g
Index
T0105
B31
5
13
s
d
g
Index
T0105
B32
5
13
g
s
d
P05
B33
5
13
d
s
g
P05
B34
5
13
s
d
g
P05
B35
5
13
d
g
s
F018
B36
5
13
d
s
g
F018

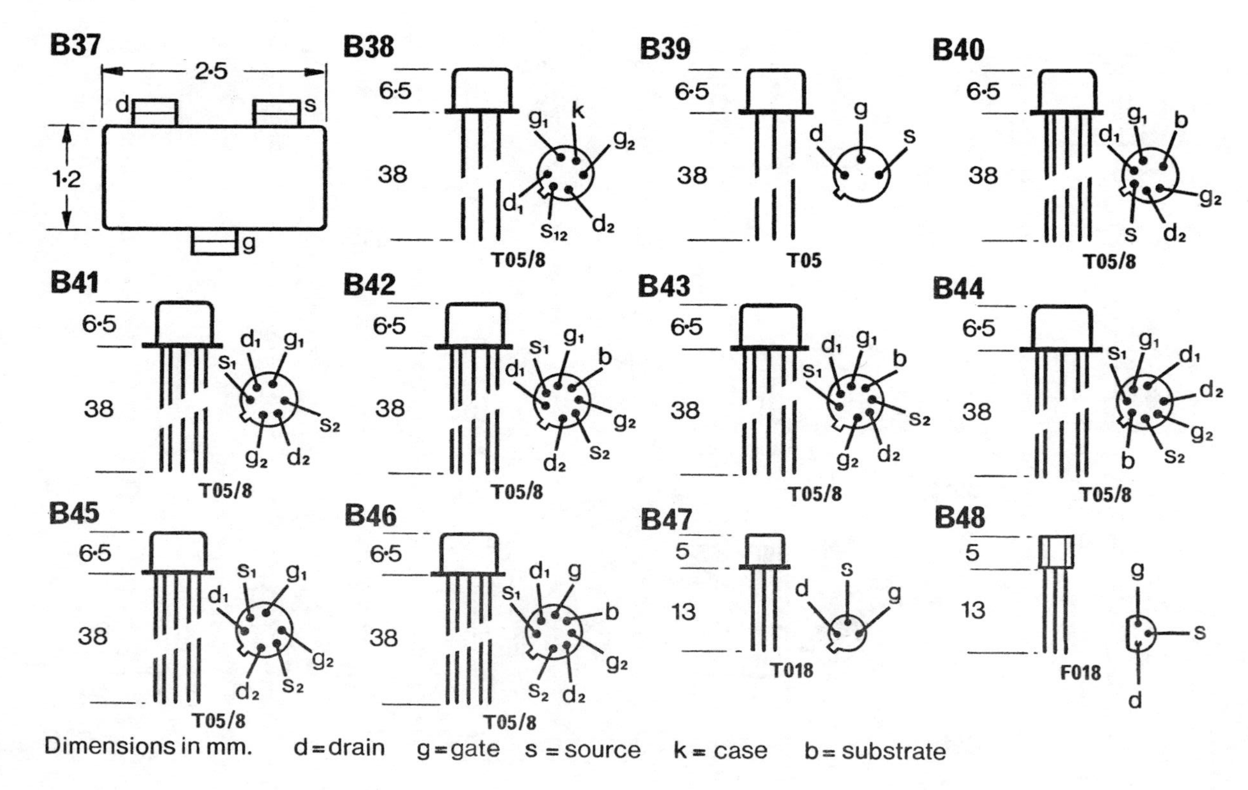

Dimensions in mm. d = drain g = gate s = source k = case b = substrate

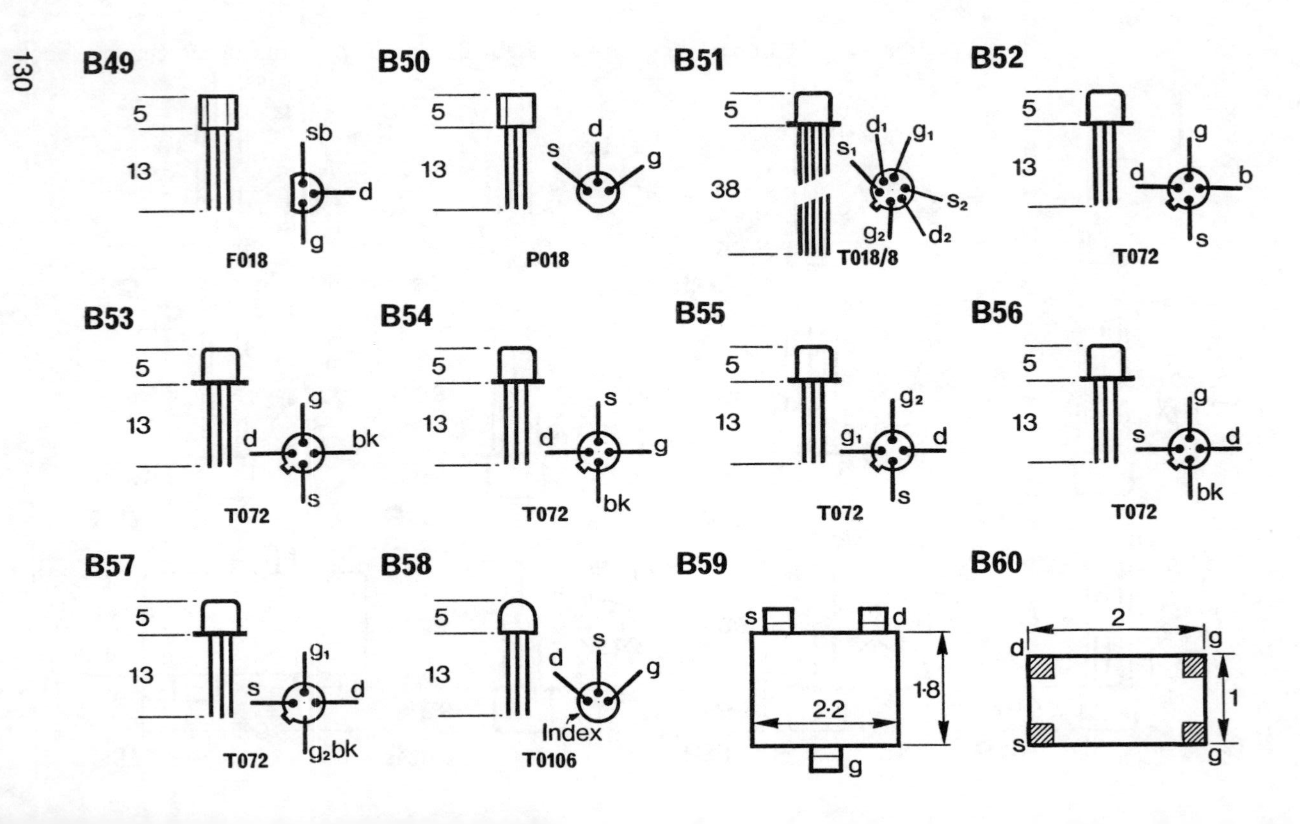
B49
5
13
sb
d
g
F018
B50
5
13
s
d
g
P018
B51
5
38
d₁
g₁
s₁
s₂
g₂
d₂
T018/8
B52
5
13
g
d
b
s
T072
B53
5
13
g
d
bk
s
T072
B54
5
13
s
d
g
bk
T072
B55
5
13
g₂
g₁
d
s
T072
B56
5
13
g
s
d
bk
T072
B57
5
13
g₁
s
d
g₂bk
T072
B58
5
13
s
d
g
Index
T0106
B59
s
d
1·8
2·2
g
B60
2
d
g
1
s
g

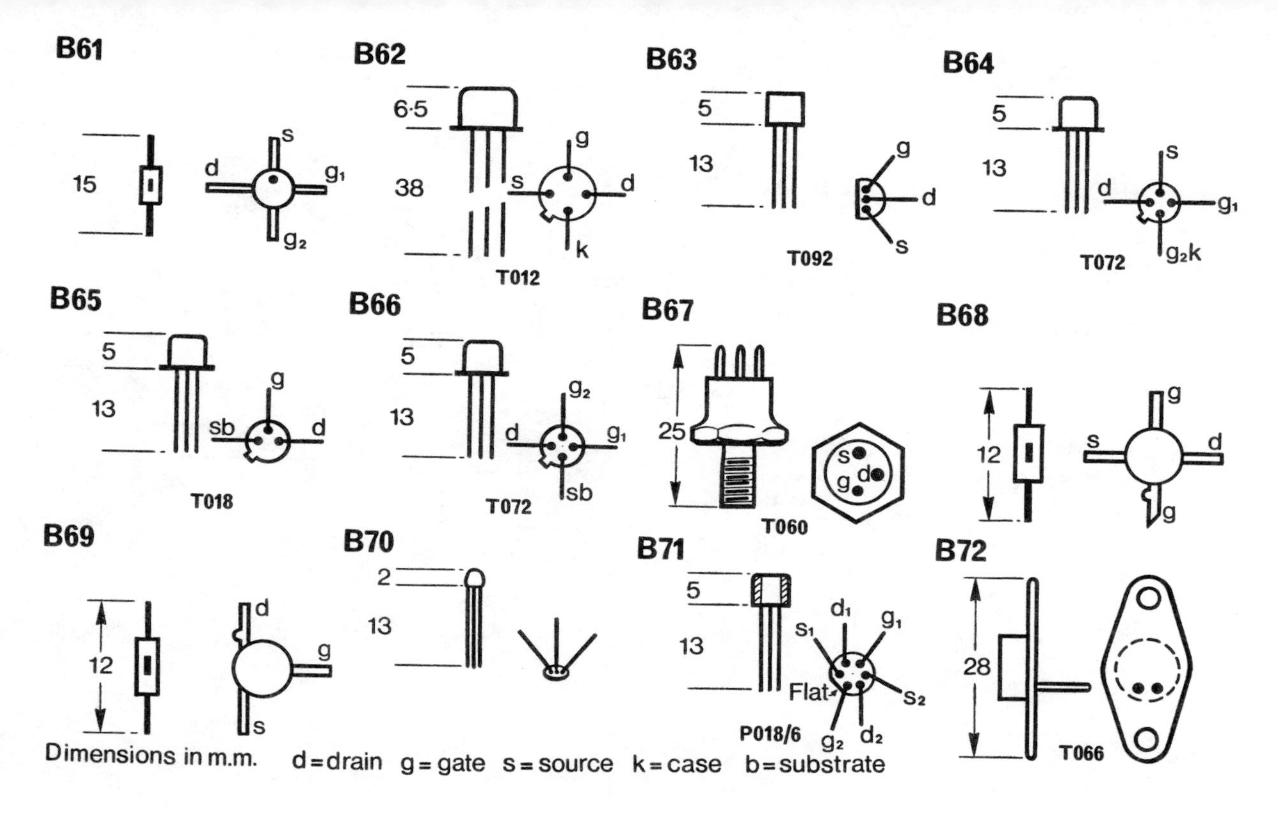
B61
15
d
s
g1
g2
B62
6·5
38
T012
g
s
d
k
B63
5
13
T092
g
d
s
B64
5
13
T072
s
d
g1
g2k
B65
5
13
T018
g
sb
d
B66
5
13
T072
g2
d
g1
sb
B67
25
T060
s
g
d
B68
12
g
s
d
g
B69
12
d
g
s
B70
2
13
B71
5
13
P018/6
d1
g1
s1
Flat
s2
g2
d2
B72
28
T066
Dimensions in m.m.
d = drain g = gate s = source k = case b = substrate

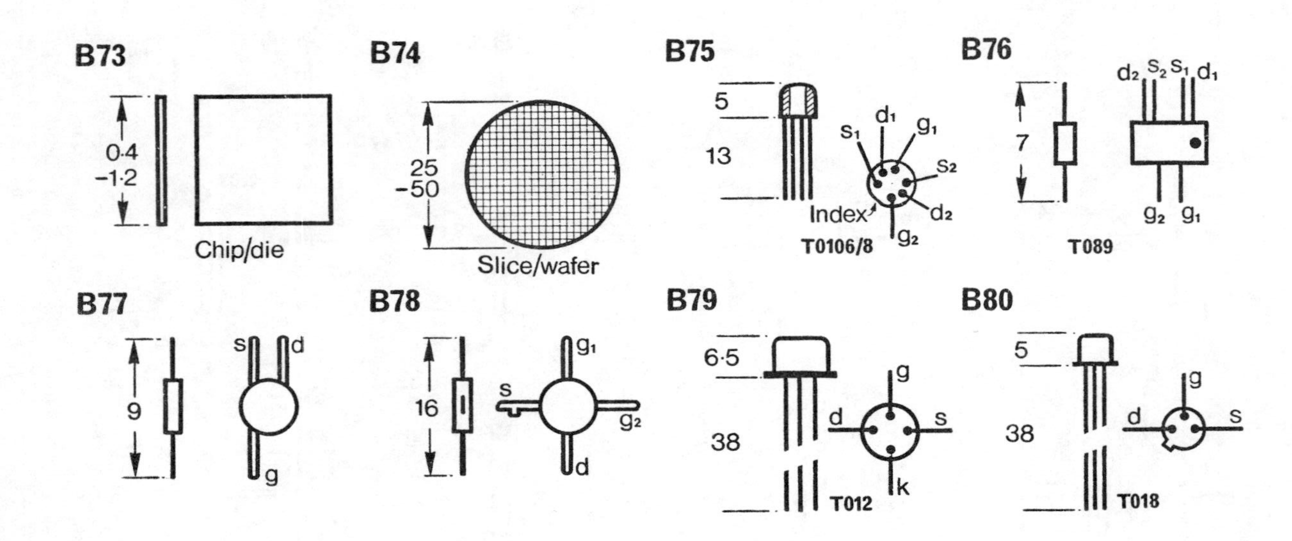
B73
0·4
-1·2
Chip/die
B74
25
-50
Slice/wafer
B75
5
13
d1
s1
g1
s2
d2
g2
Index
T0106/8
B76
7
d2 s2 s1 d1
g2 g1
T089
B77
9
s
d
g
B78
16
g1
s
g2
d
B79
6·5
38
g
d
s
k
T012
B80
5
38
g
d
s
T018

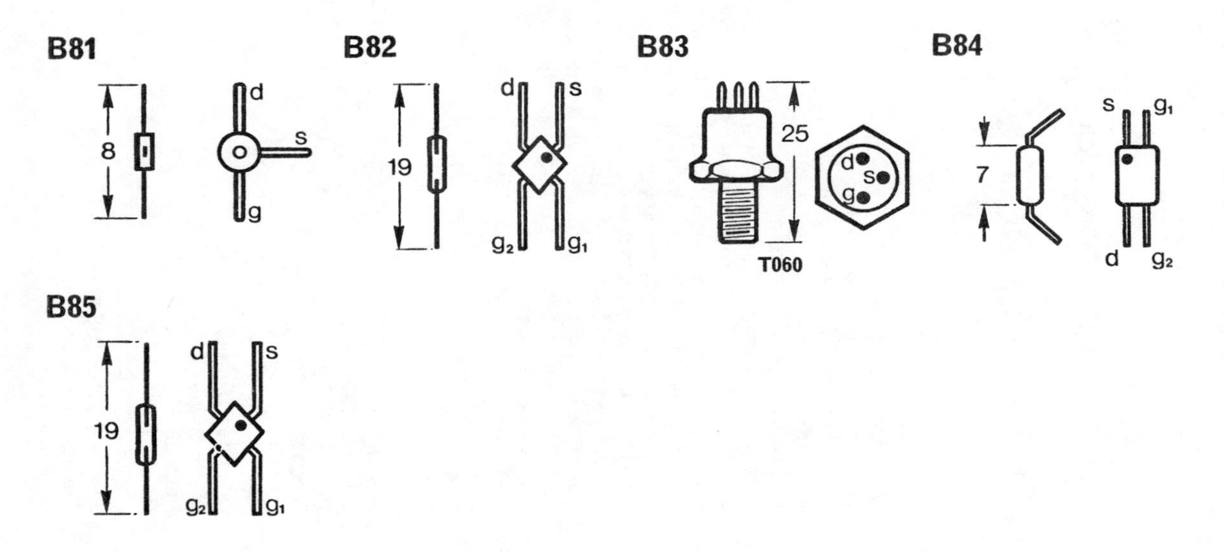

Dimensions in m.m.   d = drain   g = gate   s = source   k = safe

Appendix C

Manufacturers' House Codes

2N, 3N	USA (EIA-Jedec) Standard
2SJ, 3SJ ...	Jap (Jis) Standard, fet, p-channel
2SK, 3SK ..	Jap (Jis) Standard, fet, n-channel
A	Amperex
AD	Analog Devices
BC	Euro (Proelectron) Standard, low power, lf, consumer
BF	Euro (Proelectron) Standard, low power, hf, consumer
BFR	Euro (Proelectron) Standard, low power, hf, industrial
BFS	Euro (Proelectron) Standard, low power, hf, industrial
BFW	Euro (Proelectron) Standard, low power, hf, industrial
BFX	Euro (Proelectron) Standard, low power, hf, industrial
BSV	Euro (Proelectron) Standard, low power, switch, industrial
BSW	Euro (Proelectron) Standard, low power, switch, industrial
BSX	Euro (Proelectron) Standard, low power, switch, industrial
C	Crystalonics, Semitron
CC, CM, CP	Crystalonics
D, DP	Dickson
DN	Dickson
DNX	Dickson
DPT	TRW
DU	Intersil
E	Siliconix, Teledyne
ESM	Thomson–CSF
F	Fairchild
FE, FI	Fairchild
FF	Crystalonics
FM	Fairchild, National
FN	Raytheon (obs.)
FP	Siliconix
FT	Fairchild
G	Siliconix
GET	General Electric (USA)
GME	General Microelectronics
HA	Hughes (obs.)

HEP .... Motorola
HEPF ...... Motorola
IMF .... Intersil
IT .... Intersil
ITE .... Intersil
J .... Siliconix
JH .... Solidev
K .... KMC
KE .... Intersil
LDF .... Mullard
LS .... Ledel
M .... Siliconix, Intersil
MEF .... Microelectronics
MEM ...... General Instruments
MFE .... Motorola, metal can
MK......... Mitsubishi
MMF....... Motorola, matched pair
ML .... Plessey
MMT....... Motorola
MP .... Motorola
MPF .... Motorola, plastic
MT .... Microelectronics, Plessey
NDF .... National Semiconductors
NF .... National Semiconductors
NKT........ Newmarket
NPC .... Nucleonic Products
P .... Teledyne, Siliconix
PF .... National Semiconductors
PFN .... Dickson
PH .... Akers
PL .... Texas Instruments
S .... Akers
SC .... Philco (obsolete)
SD .... Solid State Scientific
SES .... Thomson–CSF
SFF .... Thomson–CSF
SFT .... Sescosem
SI .... Akers
SU .... Teledyne
T .... Siliconix
TIS .... Texas
TIXM ...... Texas
TIXS .... Texas
TN .... Teledyne
TP .... Teledyne
U .... Siliconix, Teledyne
UC .... Solitron
UT .... Siliconix
VCR .... Siliconix
VF .... SGS
VI .... United Aircraft
VMP .... Siliconix
WK .... Walbern
ZFT .... Ferranti
ZTX .... Ferranti

Appendix D

Manufacturer Listing

(and codings for tabulation 'Supplier' column)

AEG **AEG—Telefunken (UK) Ltd.**, Bath Road, Slough, Berks, U.K.
AKE **Akers—AME**, 3191 Hoeten, Norway
AND **Analog Devices**, Route 1 Industrial Park, PO Box 280, Norwood, Mass., U.S.A.
AMP **Amperex** Electronic Corporation, Semiconductor Division, Slatersville, Rhode Island, 02876, New York, USA
CRY **Crystalonics** Division, Teledyne Inc.
147 Sherman Street, Cambridge, Massachusetts 02140, USA
DIC **Dickson** Electronics Corporation
310 South Wells Fargo Avenue, Scottsdale, Arizona, 85252, USA
FCB **Fairchild** Semiconductor Ltd
Kingmaker House, New Barnet, Enfield Road, Hertfordshire, UK
Tel 01-440-7311
FCU **Fairchild** Semiconductor Corporation
401 Ellis Street, Mountain View, California, 94040, USA
FEB **Ferranti** Ltd
Gem Mill, Chadderton, Oldham, Lancashire, UK
Tel 061-624-6661
FUJ **Fujitsu** Ltd
1015 Kamikodanaka, Kawasaki, Japan
GEU **General Electric** Company
Semiconductor Product Department Build, 7, Electronics Park, Syracuse, New York 13201, USA
GIB **General Instruments (UK)** Ltd
Cock Lane, High Wycombe, Buckinghamshire, UK Tel High Wycombe 445311
GIU **General Instruments** Corporation
600 West John Street, Hicksville, New York, 11802, USA
HIJ **Hitachi** Ltd
Electronic Devices Division, Marunouchi Building, 4 1-Chome, Marunouchi Chiyoda-Ku, Tokyo, JAPAN
INB **Intersil** Inc.,
8 Tessa Road, Richfield Trading Estate, Reading, Berks, UK
INS **Intersil** Inc.
10900 North Tantau Avenue, Cupertino, California, 95014, USA
ITU **ITT** Semiconductors
3301 Electronics Way, West Palm Beach, Florida, 33047, USA
LED **Ledel** Semiconductor Inc.
718 N. Pastoria Avenue, Sunnyvale, California, USA
MAT **Matsushita** Electronics Corp
Saiwaicho 1 – Takatsuki, Osaka, JAPAN
MCB **Microelectronics** Ltd
York House, Empire Way, Wembley, Middlesex, HA9-OPA, UK
Tel 01-903-2721
MCE **Microelectronics** Ltd
38 Hungto Road, Kwung Tong, Kowloon, HONG KONG
MIT **Mitsubishi** Electric Corporation,
Tokyo Buildings, 2–12 Marumouchi, Chiyoda-Ku, Tokyo, JAPAN
MOB **Motorola** Semiconductors Ltd
York House, Empire Way, Wembley, Middlesex, HA9-OPA, UK
Tel 01–903–0944
MOU **Motorola** Semiconductor Products Inc
5005 East McDowell Road, Phoenix, Arizona, USA
MUB **Mullard** Ltd
Mullard House, Torrington Place, London, WC1E-7HD, UK Tel 01–580–6633

NAB **National** Semiconductors (UK) Ltd
19 Goldington Road, Bedford, UK

NAT **National** Semiconductor Corp
2900 Semiconductor Drive, Santa Clara, California, 95051, USA

NEM **Newmarket** Transistors Ltd
Exning Road, Newmarket, Suffolk, UK Tel Newmarket 3381

NIP **Nippon** Electronic Company Ltd
1753 Shimounumaba, Kawasaki City, JAPAN

NUC **Nucleonic** Products Co Inc, Nucleonic Components Devices Division
6660 Variel Av, Canoga Park, California 91303, USA

PLE **Plessey** Semiconductors
Cheney Manor, Swindon, Wilts, UK

RCB **RCA** (Great Britain) Ltd
Lincoln Way, Windmill Road, Sunbury-on-Thames, Middlesex, UK, Tel Sunbury 85511

RCU **RCA** Corporation, Solid State Division
Route 202, Somerville, New Jersey, 18876, USA

RTC **RTC** (La Radio Technique – Compelac)
130 Avenue Rollin, Paris 11, FRANCE

SEV **Semitron** Ltd
Cricklade, Wiltshire, UK Tel Cricklade 464

SGB **SGS/ATES** (UK) Ltd
Planar House, Walton Street, Aylesbury, Buckinghamshire, UK
Tel Aylesbury 5977

SGI **SGS** (Societe Generale Semiconduttori Spa)
Via C Olivetti 1, Agrate Brianza, Milan, ITALY

SIB **Siemens** (UK) Ltd
Great West House, Great West Road, Brentford, Middlesex, UK
Tel 01-568-9133

SID **Siemens** Aktiengesellschaft
Bereich Halbleiter, Balanstrasse 73, Munich 80, WEST GERMANY

SIX **Siliconix** Incorporated
2201 Laurelwood Road, Santa Clara, California, 95054, USA

SOT **Solitron** Devices Inc
1177 Blue Heron Boulevard, Riviera Beach, Florida, 33404, USA

SOU (UK Division of **Solitron**) **Solidev** Ltd
Edison Road, Bedford, MK41-0HG, UK Tel Bedford 60531

TDY **Teledyne** Semiconductor
1300 Terra Bella Av, Mountain View, Cal., USA

THS **Thomson CSF** (UK) Ltd
Ringway House, Bell Road, Daneshill, Basingstoke, Hants, UK

TIB **Texas Instruments** Ltd
Manton Lane, Bedford, UK Tel Bedford 67466

TIU **Texas Instruments** Inc
Components Group, 13500 North Central Expressway, PO Box 5012, Dallas, Texas, 75222, USA

TIW **Texas Instruments Deutschland** Gmbh
8050 Freising, Haggertystrasse 1 Freising, WEST GERMANY

TOB **Toshiba** (UK) Ltd
Toshiba House, Great South West Road, Feltham, Middlesex, UK

TOS **Toshiba** (Tokyo Shibaura Electric Co)
Toshiba Building, 2–1 Ginza, 5-Chome Chou-ku, Tokyo, JAPAN

Notes

Notes

Notes